Chrology

Science of All Sciences
Unification of All Knowledge

by Ulrich Ndilira Rotam

DORRANCE
PUBLISHING CO
EST. 1920
PITTSBURGH, PENNSYLVANIA 15238

The contents of this work, including, but not limited to, the accuracy of events, people, and places depicted; opinions expressed; permission to use previously published materials included; and any advice given or actions advocated are solely the responsibility of the author, who assumes all liability for said work and indemnifies the publisher against any claims stemming from publication of the work.

Dorrance Publishing Co
585 Alpha Drive
Pittsburgh, PA 15238
Visit our website at *www.dorrancebookstore.com*

ISBN: 978-1-6453-0217-9
eISBN: 978-1-6453-0843-0

This book is dedicated especially to my lovely father
Ndilira Ndilingaye Soyos.

Contents

Foreword

My name is Ulrich Ndilira Rotam; I am a researcher in physics and cosmology. I conducted my research on the different parts of this book, which deal with the universe in its immense extent, in a very solitary and complicated field for over eighteen years of inquiry.

The research has been conducted in a generalized way on several domains to understand if there is a single law that governs all sciences, all literary studies, our existence, and all our knowledge on different generalities in a single model.

This model explains all the different areas of presence in the universe, within in the face of all the studies that we have undertaken and acceded to throughout human progress.

This research and study have led me to discover a simple and absolute law in its originality that governs the presence of all existence in the universe in a complex way according to the space, existence, time, and our scalable factors.

To carry out this research, I spent a very difficult time in my life looking for a path that should lead me on a simple, free, and deep way to safely manage and conduct my inquiries, quests but the different systems already on site do not help me let go of my way and my curiosity.

Finally, it's a great fight that I led all these years alone. If I let myself go to a specialized study, I would never arrive on this complex road and I would be lost because the universe never gives us its secrets on a specialized study but rather gives us a thin clarification on a small part of everything his secret is carrying.

All the efforts I have made, with a sincere curiosity and a deep envy, in the respect of all those who exist, seen, studied, in our world have thirsted me to go through and seek to understand the substance of all the studies described by man, and try to measure if there is a single order that exists, and which must govern everything in the face of the realities of the whole universe.

Consequences, these guided me on an exaggeratedly complicated and extraordinary road. All my life has been concerned with more works of calculations, and extensive research work, very complicated to understand the basics of each scientific study, the basics of the different fields of literary studies, and religious that we have so far.

Those took me, to go through and dig deeply several works and phenomena on different eras of our eras of existences, and evolutions with serious questions, to know more face the original realities, the universe possess in its immense way.

I finally went through several documents and several works of great scientists, great geniuses, great writers, great philosophers, great chemists and physicists, great mathematicians, great universal conventions, laws of domination, different scientific disciplines artistic, literary, religious etc. to conduct my quests.

The pressure is very difficult to be free and enter this broad field of cosmological visions to understand it, with a simple law. I finally created a tunnel in which, I feel free and I allowed myself to go further, with the thought, and the mathematical tools, in the different confines, of all our different systems and existences in the universe which are governed by very complex natural laws and very simple natural laws.

All these steps have finally opened to me and access to a door of clear and simple comprehension, of the whole universe in its gigantic whole and in its deep tiny confines, described by **CHROLOGY**. Several works that have been undertaken and carried out by great researchers, great scientists, and great geniuses of our planet, at different times; have allowed me to advance in these extraordinary quests of research.

I thank all humanity any intellectual purpose and especially all those who contributed to a more exceptional reflection on our world, and on the different parts of the universe, that this great genius of all the geniuses, the absolute designer of this extraordinary system of everything, of all the time, of the whole space, of all existence, and all presences allow us to reach.

In my lonely freedom, inside this tunnel containing more gigantic roads, more tiny, more complex, extraordinary, absurd, and even scary. I was very scared and I was ready to abandon my work several times, but especially the efforts and the courage to continue in my deepest curiosity as a human being, helped me a lot not to give up, but to continue to go further in my research, to corroborate several hypotheses with new techniques of apprehension and new visions more natural and original in this tunnel configures very near or very far, for the quest for a single and simple law that describes the whole universe in a whole, facing our different disciplines and our different evolutions that the physics sciences and other disciplines are unable to answer etc.

I wanted in this book, and the editions that will follow to show and to share with my readers, all those that this complex system presents us and allows us to access according to our evolution with our intelligence, our different properties, formulas, law, axioms, conventions, efforts, etc. according to original realities, to be able to understand it and access it without modifying it.

All these studies show exactly how all those who are material or existences must appear in the whole universe. Being the result of a very broad and very complicated research path, these results are named in a new name: **CHROLOGY**.

CHROLOGY is the typical model of vision, of understandings, of original descriptions of the whole universe and its systems, facing all our intelligence and our K factors of existence.

CHROLOGY is the theory of the unification of all human knowledge, the physics of all physics, the literature of all literatures, and so on in a unique model, to represent and understand in an original way the whole universe in its different faces of existences and its different synchronizations, on all the geometry of its dimensional structure.

These results of research and studies are not to vex, or to thwart the world. But it is rather to contribute and explain with new methods, all those that the entire system of the universe allows us to access, understand, and use to the limit that escapes our standard methods.

To clarify this new vision to my readers, we must understand that we are only a small part, of another part, that exists in several other parts of this wonder at the tiny scale, or at the gigantic scale that the designer has performed with immeasurable beauty to our scale.

To read about this entire complex expanse in every sense in the face of our intelligence and our **K factor** of **existence**, we are very insignificant with our progress and our knowledge, to detect the complete mystery of the whole universe.

I have spent more than 18 years looking, studying, and investigating all our progress and know-how in every way to compare it to how much percent we are for the quest for a simple theory or law, which should serve us in the static and total understanding of the universe. But the result is very scary, and very far from those we expected.

Writing a book or producing the results of in-depth research of the absolute understanding of the universe that is seen and understood from different angles and different doctrines is a titanic work, very complicated, and difficult with several contradictions but the technology of the internet helped me a lot in this huge investigation during the past ten (10) years in my quest.

This book opens us a new way and a new conception to add to all our knowledge to realize and understand how the absolute mystery, the immeasurable beauties and the complexities, of the whole universe must present itself to our access with a unique law.

This book is also presented in its unique model, with several new terms, introducing the facts and phenomena, which were missing from our standard models, to describe and explain the universe.

We must focus in his deep explanations, to fully understand the novelty that he presents under its different principles on the existence of domains, which is available to us for its existence described by **CHROLOGY**, to help us advance in the quest of deep explanations of all its existence and perceptions, which were missing in our **languages**, our **formulas**, our technical progress etc.

Without however in any way, if the different representations and in-depth explanations of Chrology in this book and the subsequent editions, demonstrating the consequences of representation and complex understanding, of the whole universe with some of its aspects of which we sum unable to explain all its simplicity, can hurt or touch someone, an organization, a doctrine etc. in their world of understandings, we apologize deeply, because the desire that being human to seek to understand the mystery of the universe, and the many questions that elude us is a multiple and profound cause of the realities that must touch us.

Several editions and several conferences will follow this first part to develop and clarify the deep meanings of Chrology to the public in its original aspect.

Chrology will completely change the way we view nature and all our achievements with new methods, which will allow access to many discoveries and innovations.

>*<<We live on a very simple model that appears flexiblyonits five worlds to deeply confuse us to his full understanding and his origin if we are in one of these complex presences.>>*Ulrich Ndilira Rotam

Introduction

In the concepts currently admitted, scientific progress is the key that allows us to detect certain mysteries, certain truths, and certain absolute enigmas of the universe, opposite to our human thoughts, with great rigors of logic, calculations, and truths at the scale of our understandings present in many parts of our accessible universe.

In our universal human realities to access and understand the different phenomena, the different constituents, which form and govern the whole universe, we must show them with the physical sciences, mathematics, and so on. Using several parameters in nature, which we ourselves are part of and at the end all must be visible, concrete, logical, real, and especially all our sense organs allow us to grasp, feel, and recognize the facts before our world accepts, these descriptive ideas and optimizes them in the classical concepts of our visual and real cosmos, to use them for the quest for absolute understandings of the different mysteries present in its complex set which we will describe in Chrology.

It is the absolute knowledge of phenomena and existences of the facts of the universe obeying laws and verified by our experimental methods. But the whole universe is described by very complicated laws and by very easy laws, which some of our experimental methods are unable to explain.

Every existence in the universe is governed by a law of evolution. Time and space contain and accompany every system towards its evolution. But from the smallest to the most gigantic system, in perpetual evolution, no one has yet to allow and to understand a complete system that shows all the secrets of the universe.

Our existence and simple understanding in the universe allow us to perceive the world in many ways after passing vision and other senses to the brain for interpretation. But on the complexity of the whole scale of understanding the universe, in our fields of vision and other sense organs, in relation to our brains, our computers, and the instrumental materials that we have designed to obtain, analyze, and verify results, then to classify in our human knowledge without however modifying the concepts, on all its originality, constitute an abstract representation and also a very complicated answer to correctly interpreted because we are included in this part and each function of the time elapses takes us on other mysteries discovered and we continue riddles from enigmas to enigmas.

Therefore, all our knowledge bases are not free, and we evolve challenges into disputes, from one group to another, from one doctrine to another, from one society to another, from one religion to another, from one race to another, from one interest to another, and so on.

>><<*The ones I look at are not exactly the ones you are looking at, only the universal conventions allow us to see and say the same things.*>> Ulrich Ndilira Rotam.

We are very far from the deep and absolute understanding of all the systems that exist throughout the cosmos.

Man possesses his brain and his sense organs, with the different devices of visions, measurements, studies, etc. and the mathematical formulas, allowing us to look at the simple or complex compounds present in the universe and the artificial products, in our environment to describe them on the scale of our knowledges with the different exact and applied sciences that we possess.

All these methods evolve according to our existence and there will be new techniques, in the future better than in the past and the present because everything evolves.

These methods and techniques allow us to take a very small step towards an absolute quest for total understanding of our existence and absolute knowledge of the entire universe.

The **universe** studies with several scientific disciplines with a unique logic, but very diversified and complex in its formulations.

If we are in the world of the tiny or in the world of macros, the different studies are done under different principles but to adopt it, in the environment that has been attributed to it by nature.

The only order that exists is that each system is founded under a single base, in its world irreversibly by generating actions that give rise to other systems, which engenders yet other actions that give other systems, up to the state where one apprehends to study it.

> *<<The great universal laws are simple to detect but it is the universal path that we have undertaken that takes us away from its capture and we must undertake many efforts to arrive at a simple concrete result.>>* Ulrich Ndilira Rotam.

Looking at all the accessible elements that make up the universe with our eyes and vision devices of different technologies, we can classify them into two groups:

- **Visible**
- **Invisible**

These groups can be seen with **five (5)** known dimensions on the set of **twenty-one (21)** for the whole universe described by Chrology:

- ○ Minimal (**with length, width, height**),
- ○ Extra Small (**with length, width, height**),
- ○ Large (**with length, width, height**),
- ○ Extra-large (**with length, width, height**),

- o Normal (**with length, width, height**).

In the invisible group there are two appearances:

- **Existence**
- **Non-existence**

In more detail, Non-existence also contains two points:

- **Physical presence**
- **Physical absence**

The physical presence is an approach of a material existence (natural or artificial) that can be interpreted with laws, relations, properties, formulas, axioms, etc. in physics, mathematics, biology, and chemistry etc.; accessible to man.

Physical absence is an approach to a null existence, of which no material presence (**natural or artificial**) can be recorded and interpreted by laws, relations, properties, formulas, axioms, etc. in physics, mathematics, chemistry and biology etc.

All these elements are interpreted by our consciences and intelligences, in the face of the realities that the whole cosmos presents to us and allow to access without modifying it.

It is in this way that our existence allows us to access and understand, with the help of our sense organs, our instruments, and our weapons of mathematical calculations centralized with our brain, to analyze several phenomena of nature and the daily problems etc. But so far there is a serious problem: many facts and mysteries in the universe, others gathered around us, others far in our fields of vision, others completely out of access to our world etc. we escape in our conventions that we do not have the means and the total access for the detected ones mastered and checked them to understand with certainty the real answers to some of its questions:

- **Why?**
- **How?**
- **When?**
- **Where?**

In the existence of Cosmic Mosses, parallel bubble universes, superclusters of galaxies, clusters of galaxies, galaxies, our solar system, the Sun, our presence, our eyes, our tiny objects, molecules, atoms, quarks, universal interactions of the quantum void, the theory of general relativities, string theories, theories of cosmic inflation, the big bang theory, the theory of dark matter etc.

- **Do we know and really know a great thing about the whole system?**
- **Are we advanced enough in the quest to solve the mystery of the universe in its absolute understanding?**

Those are colossal and exorbitant questions that have disturbed me since my teenage years and that have led me to continue searching in this complex quest to create a tunnel in which I feel free and I allowed myself to get started and to go further in my deep investigations, on the different borders of the total universe, which is governed by a single law containing multiple principles, very complex, and singular principles very simple with the thought, the mathematical and scientific tools, accessible for find explanations that the universe holds in its originality, so that we can access a unique model.

These took me through my research and developed a huge and gigantic work in Theory of **Chrolos** called **Chrology** or **Physics of Chrology**.

Chrology or **ChrologicPhysic** is much wider enclosing studies in the entire system of the visible and invisible universe, to explain the real phenomena of nature without modifying it. It is the absolute knowledge in its originality of all the Cosmos as a whole.

In my solitary freedom on this road, it has been more gigantic, complex, tiny, and even boring journey to go further in my research of solutions and understanding of **n** hypotheses, **n** questions, **n** mysteries etc. that universe presents us with a simple law to describe all the laws of science in its simplicity.

Chrology or **Chrologic Physics** is the set of all theories, formulas, knowledge, discoveries, and laws of all sciences in the universe.

SUBSTRA is single concept showing the different hidden faces of the whole system known on all the dimensions of the whole universe.

Everything comes from the substra in Chrology and Substra is the composition of the whole system in all its set traditionally called atoms.

My paths in this tunnel, configured with different near and far access in the universe are studies to open other basis of thinking, for the quest for advancement and understanding of the whole universe in a single and unique concept.

This very difficult and very complicated research that I spent most of my life on, utilizes all our knowledge, technical possibilities, and facts in the universe that we have accessed so far, to complete the other work to perform in our admitted world. These works also produce a new vision in support, to show those that this complex system presents us and allows us to access, according to our evolution our intelligence, our efforts, our means by considering the different factors of realities original for to be able to access it.

Being the result of a very broad and very complicated research path, the **CROLOGY** is the result of a typical model of these natural descriptions of all these systems, facing all our intelligence and existence.

Two (2) terms: One called **Global World (GW)** and second called **Chrologic World (CW)** are determined and used throughout all Chrology and present as one of the bases for the understandings of the whole system of the scientific world and especially the immeasurable beauties and knowledge that nature allows us to access in its entire entirety.

The Global World (GW) and the Chrologic World (CW) will be developed in the scientific part of Chrology study.

I ask my readers to ask themselves a profound question about each entity that exists; there will be several answers that we have not admitted on the scale of our understanding of the real existence without modification of all the original aspect of the system.

This book allows us to go further in our different areas of research, vision, and understanding more forward of the whole universe, in its gigantic or minimal peculiarity with its different original laws.

Several words are missing to describe certain facts, limits, and existing vision. We will use the Chrology to fully explain the results of this research.

This first edition also takes us on extraordinary questions, on the absolute foundation of all existence in the universe, with the riddles that it presents to us and which surpasses us, on the meaning of our presence, our destination and the different possibilities. The Universal Quantized.

This edition opens to readers, new models of vision, comprehension, and absolute reflections of the whole universe, unifying all knowledge in the different principles of Chrology.

This edition also makes it possible to answer the extraordinary questions, in the entire universe that we seek and continue to pose during our different periods of life, with the different scientists, researchers, geniuses, writers, etc...

Finally, Chrology opens us completely, a new way of understanding the whole universe, unified with all our knowledge.

Chapter 1

The Great Human Questions of All Time

Content

1. The real questions and challenges of our great ideas.
2. Compared Human knowledge on the scale of all existences.
3. Is our knowledge able to unveil all the secrets of the whole universe?
4. The sacred questions of all time.

The real questions and challenges of our great ideas.

Human knowledge is evolving, **the universe** is a complex that is very difficult to understand and master, many great ideas that have been accepted and used in good times to revolutionize everyday life have been contested after several eras or man are always evolving and reviewing our great intellectual progress with new means. This is the example of several cases in our scientific communities, literary, religious, techniques, etc.

This simple contradictory thought of a formula, a discovery, an invention, a product, etc. from one era to another pushes us to ask several questions is what really what is right is true in the original form of the whole universe as a whole?

Are our organizations that certify natural concepts are correct to certify these concepts in the original universe concept?

All these questions push me to ask several more pertinent questions.

Man is an element of the universe, and the universe in all its originality has several known and unknown elements with completely different functions known and unknown.

Can an element of a thing or an element of a group determine or know the whole of its group?

Can an element of a thing or an element of a group face its position and its **original presence** know the direction of evolution and the why of its entire whole? And ask the question, how was this group born?

These thoughts allowed me with some very simple consequences to understand something very big.

At our **scale** we have issued the big bang theory that explains the birth of the universe, its evolution and its end, but this theory is unable to explain those they had before the big bang.

Through the evolution and the progress of our ancestors on different times, the questions to detect and understand the universe were active as in our present time but the means are not the same. They emitted and understood the universe with several different theories which limited them considerably until the century of the light with the invention of the telescopes and microscopes to reach certain enigmas of the nature. But we must also understand that the law of nature does not change but is transformed.

The questions about the true foundations of the bases of our different disciplines on the planet earth allow us to understand very little about the original traces that the whole universe allows us to access.

Therefore, we discover, innovate, and expand our knowledges with several theories that some are valid through the recognitions and implications of these disciplines. But a very relevant question that pushes me to attack is the question of why we accept some ideas and reject others if only after those who accept such ideas will be rejected with time etc.

The institutions that can validate this knowledge are so truly true in considering all the original aspect of the whole universe to show us and accept those truths that will be absolute by considering all the details of the whole universe as a whole?

These questions remain **unanswered** because we are minuscule in comparison the universe; our lifetime on Earth is extremely short and small in power to be able to make a study at any scale to fully encompass the whole universe.

In the **first (1st) principle of Chorology**, our world where we are being and live are well defined to help us measure our knowledge and progress.

In the standard method used today we must remember that all our theories, progress, materials, knowledge, and products, will always question if an institution, organization, or person really tackles the problem at its root.

All these consequences are the exact facts of such that has placed us in a very petty and confused world of complication existence on a human scale that the whole universe allows us to access.

To tackle questions like:

- What is the true meaning of our existence?
- What are we really?
- What is the distance of the whole universe?
- What is there before and outside the universe?
- What is the last tiny piece that must exist in the whole universe?
- What is the shape of the greatest existence in the whole universe? etc.

All must involve everyone, scientists, literary, religious, public, anyone in his workshop etc. to help us understand some percentage on our scale and in our world because we are all parts of the whole universe and the universe allows and confit to each piece that we are one of its secrets that unfortunately humanity does not exploit.

Compared human knowledge on the scale of all existences.
All our efforts of knowledge of the universe of today will be very tiny and absurd in a **Chrologic future**.

Many of our research and theory in the scientific community have been accepted by one group, challenged by another, or accepted or rejected by another group so if these theories and discoveries are true?

Are these truths true to the original state of the whole set in the universe? Consequently, we advance at a very small piece compare to the complexity of our entire existence and know in relation to the original state of everything.

The more we compare, or we revise the system of life with our technologies one realizes that past eras had not had the luxuries on these technologies compare to our present time, and if we let time elapse the children of tomorrow will find our technologies too late. All these tendencies of apprehension are truly the absolute access to our evolution in the universe.

Time and **space** let us differentiate all our habits, our progress, our knowledge, etc. But especially in evolving in the universe is what is true and what real meaning with material evidence to corroborate true states to the absolute scale of these things in their true states of existence throughout the universe? These are relevant questions that no one has yet to check.

That's why we should be very careful about all our habits because we are admitted into a small part that is a part of another part up to a large part, we will never be able to understand this mystery to the absolute scale.

To describe and show with an absolute sense of our existence in the entire universe several factors will take us to not know.

Consequences that we limited considerably to the true answers of our existences and our truths are not all true on all the scale of the universe, we are confused to a Chrologic problem that it is necessary to attack the representation of all the dimensions of existence of the whole universe before answering several puzzles.

Chrology will help us to understand these existences and the ways in which their models represent themselves in the original concepts of existences. It also shows its deep concepts of any of existences.

Is our knowledge able to unveil all the secrets of the whole universe?
With the exact and applied sciences, natural sciences, engineering sciences, technological sciences, medical sciences, sanitary sciences, agricultural sciences, social sciences, human sciences, etc. we have made considerable efforts to compare to the state of our different era of existence.

With our knowledge, we always give meaning to all those we see and access in the universe and explain the real state and true to the scale of our understanding but the constitution of all of its balance and versatility or the irreversibility of all events, all existences, all evolutions in its overall state show us that we are still far from the absolute comprehension of all that we describe on our universe to its original state.

Religions give very simple explanations that we must believe without asking questions, the different literatures philosophy decry while opening several basic questions with non-tangible evidence of facts, but some fundamental things are missing to all our human progress to explain all the different synchronizations, all the different interminable beauties that all these tiny existences, all these titanic appearances, all these colossal presences, and all these phenomenal diversities that everything of the universe possesses as a whole.

He holds another secret that needs completely different means by exploiting our current knowledge for apprehended and understand it in an original and absolute way.

If we revisit, seriously, our knowledge and re-study everything there is another absolute question that is not answered.

We do not yet understand at a large percentage the true meaning of our absolute existence in the whole universe.

We live a very short time and we disappear definitively to contribute to the evolution of the whole universe.

The human body is a Chrological set. Each organ is designed to meet in its world has very important functions and well determine.

We are composed with a complex assemblage of several organs that determine us in a micro and macro world according to our Chrologic scale and each source of these organs pose another completely different question in our process of understanding our births.

Comparing ourselves throughout the universe is an abstract question because we are only a thin of another part existing overall.

The universe is a complex of bodies, elements, particles, etc. each of which plays a very fundamental role for the whole to exist, evolve, and live.

Therefore, we do not understand much of the absolute secret of the universe while we attempt with different means to explain it with our limited senses, intellectual pride, arrogance, intellectual, and materials. All that express nothing in the deep sense of knowing the whole universe?

The sacred questions of all time.

All the **science** and **literature** of the past and of our time have classified themselves in very broad and very small domains, of which the great scientists of the past and nowadays have great curiosity to detect all the great mysteries of the whole universe to understand it and explain its facts but the system is still very complicated that we need other means of research more sophisticated, to our understandings to go very far to conquer the answers to these many sacred questions, like our black beak which some are valid and some almost invalid:

- Where do the origins of the universe come from before the Big Bang?
- What was there before the appearance of the universe? What are its concepts? Its forms?
- How are they born? What are the true secrets of its nature?
- Why a physical presence in the universe? What will be its final limits after the Big Bang? What will be its final limits after the quantum world?
- Where do we go with all those around us?
- What is the real meaning of our existence?
- What is the real meaning of death?
- Why do we live and disappear?
- For how long will the land exist?
- Is there any other possibility for particles without mass to go faster than light?
- Are there any things in the over up of the whole universe?
- How, and why, the systems of galaxies and planets have positioned themselves in the universe?
- The MAN in all the cosmos represents what?
- What are the meanings of our contributions to the evolution of the whole system?
- Is there any other human outside of the earth in the universe system?
- What does it represent and explain our situation on earth in the whole system of the cosmos?
- Why does the egg and sperm merge only in the microscopic world and grow to give life to our scale?

- Why do men kill each other for material reasons?
- Why do we hurt ourselves, kill each other with our sentimental reasons? Racial? Religious? Radical?
- Why do we hate other nations?
- Why do we think to be superior to others?
- Why does money change humans?
- Why do not we live forever?
- Why do we die in many ways?
- Why are we not comparing anything to the scale of existence of the entire universe?
- Why does human knowledge never stop?
- Are our technologies enough and powerful to unveil all existences of the universe?
- Does antimatter exist or is it an outdated concept?
- What secret holds dark matter?
- What are the real limits of the horizons of the cosmos throughout his system?
- Are all our conventions built by man enough to govern the earth?
- Why are the dominations of nations over other nations not forever?
- What is the formula for making life on earth happy?
- Why are we obese and jealous of other people? Other institutions, other ideas?
- Why the existence of the tiny world?
- Why the presence of the macro-cosmic world?
- What existences are before the tiny world and what presences before the macro-cosmic world?
- Do we know really ourselves? And if we do, why we are present now? Etc.

It is to all these great questions that the sciences, literary studies, and all people on earth who think have always tried to answer that other validate, other invalidate, and still tries to answer several riddles of the universe in its different dimensions of existence.

It is also these exorbitant questions that completely upset me and pushed me into an immeasurable inquiry into the quest for some answers and ended up opening the birth of Chrology.

Chapter 2

Our Limits of Existences and Understandings in the Universe

Content

1. Our existence is born and excluded from many means to describe the whole universe.
2. All that exists can be conceived artificially in the universe.
3. Our presence in the universe is like a millisecond or billion years compared to all its different absolute scales.

Our existence is born and excluded from many means to describe the whole universe. Space, time, and the K existence factor are real parameters of presence throughout the universe to surely confuse us with the true foundations and the true face of the deep understanding of the whole universe with respect to our scale and existence.

We are considerably limited and absolutely excluded with other methods, other materials, other technologies, and some **mathematical** expression to describe the whole universe in a simple concept to understand all its secrets.

The place where we live is **homogeneous and favorable** to our existence to apprehend, to identify, to study, to understand, and to explain the various facts present on the universe that this same universe allows us to reach with our means and **technologies** through our different times.

But those who escape us are the staggered quantification of its entire ensemble, with the laws and visions for the understanding above these two domains to name in the first principle of **Chrology**: **Extra micro** and **Extra macro**.

Evolution is a natural parameter coming from the **SUBSTRA** described by **Chrology** according to our human technology and as a function of time.

Chrology explains the evolution of the Cosmos in its natural and original aspect with several approaches of its different **mini and macro horizons**.

The principles of composition of a matter in the universe let us reveal a small part of its secret in the micro world of its atoms, its electrons of its **elementary particles** that we know with chemistry and quantum physics.

But matter itself, which is an energy, leaves us its presence in several forms that we must not confuse these different forms with the different particularities for which exists in its originality as a presence and an entity in its domain in the universe where she must introduce herself.

The determinations and understanding of Chrology are a comparative law on different scales of the entire universe from finer to grander.

Imagine a sample of these two divisions Chrologic in a MACHRO or NORMA world that we access then we will be able to detect a great mystery of the cosmos and eliminate our doubts about our different theories and absolute understandings.

Chrology allows for a deeper study of all existences in the cosmos according to the advances of our intelligence and our progress.

It opens a door to other mathematical factors that help us to understand many existences in the universe.

Let's do an experiment taking three positions of existence in the universe:

1. The first point of positioning on the top of the status of freedom.
2. The second point of positioning on the planet Saturn.
3. The third point of positioning in a sample world of microbes.

Consequences we will never manage to understand the complete mystery of the universe and on our three points of positioning we will always be limited and confused on the absolute comprehension of our existence we will understand the universe in an abstract and limited way.

The relevant question is that we have changed the aspect of the original meaning of our existence into a single model where we mechanically follow the process of making money and putting ourselves with our pride in action to say that we are most important on the whole cosmos if and only if we understood how the universe in its tiny and gigantic appearance gives us our sense of existence as the pieces that contribute to its evolution and represents us in a very tiny aspect, we must ask ourselves another question that many of our facts are nothing and there is another simple formula that exists that should make us happier on earth for ourselves, from our families to families, from our organizations to organizations, from country to country, from continents to continents.

These fabulous contains will eliminate our pride, our vengeance, our crimes, our malice, our discontent, our ignorance, our protests, our despair, our bigotry, our various discrimination, our hypocrisies, our lies, our falsehoods to build a safe world...

All these tendencies are the proof that our original liberality was completely diverted to take on another meaning, of evolution, because the system itself uses us towards different senses in the Cosmos even if the whole universe does not give us all its secret to understand it.

Our evolution is a very tiny aspect with a very mean time and space compare to the gigantic scale of the universe and the principles of Chrology shows the whole universe to a meaning of evolution, which is not positive, not negative, not neutral, but with a qualification that we lack in our conventions compared to all our knowledge.

Experience 1

Let's experiment by studying the physical aspect of two completely different worlds: a more macro world and another more micro world.

Consequences in classical studies we have issued several explanations in sciences of miniatures and macro sciences, but in Chrology no one can understand that the two worlds are irreversible in all.

Facts

On the study of **movements**, the ways in which atoms are formed to give molecules or materials, the positioning of planets, galaxies, the micro space, the macro space, etc. are all irreversible so that the physical aspect of these two worlds exists in their originality.

Bacteria or viruses exist in their worlds with a perceivable physical aspect so that we can access their presence with a K factor of evolution in its original existence.

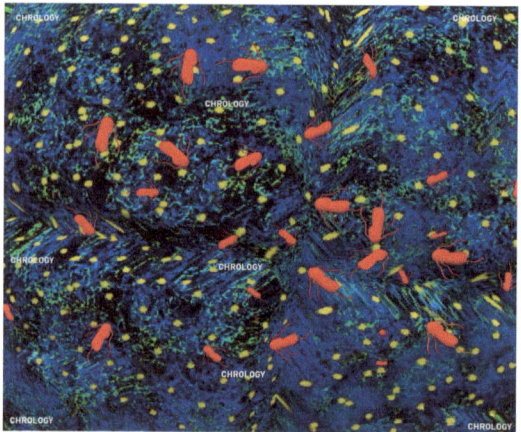

Chrology Picture 1

Example

Take the microbes that accumulate on a large portion of a sugar paw as draw in: Chrology Picture 1 below.

In the world of these microbes they are all governed by a universal law that makes them live in their space and time.

They are all confused for a better way of life for them, but it will be very difficult for them to understand if our world exists with our astronomical scales, our planes, our technological progress our life our journey, our computers, our weapons, our boats, our airplanes, our conferences, our oceans, our mountains, our rivers, our concerts, soccer world cup, our Olympic games, etc. who are no longer at their doors.

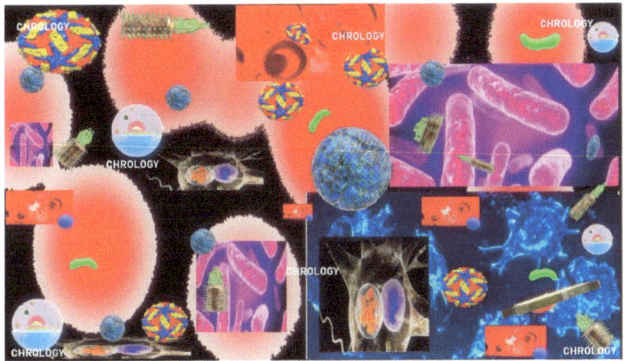

Chrology Picture 2

They can take all their techniques, theories, etc. they possess. But they will never be able to understand the world at the scales where we live. If we mechanically cut and destroy a portion of this sweet paw, the microbes will be on a completely different logic and will eventually be a material portion that after several natural factor interventions will become another thing that would contribute to the system.

This example is the same for all existences in the whole universe. We are limits in our world of logic that can be applied in all space and time that is for us but very difficult to come to understand this wide world that is beyond our space and time, or this thin world that and outside our space and time.

If we apply Chrology we will understand on a very large scale and on a very thin scale the whole universe.

The planets and galaxies also exist in their worlds with a physical aspect of another tendency so that we can access their presence with a K factor of evolution and original existence. See this example of large scale of world draw in Chrology Picture 3.

Chrology Picture 3

We live in a very limited world for the absolute comprehension of the whole universe in its originality.

Time and space emit an attachment to existence and presence in an original and complex way to appear to us through our different technology.

In Microscopic World:

The Bacteria in Microbiology with a K Chrological factor of evolution and original existence to access them with one of Chrologic principle.

The Galaxies in Machrology with a K Factor of evolution and original existence to access them with one of Chrologic principle.

Let's take another very simple example of football, to clarify the understanding of a Chrologic evolution:

Football in October 1848 was the year of the first attempt to unify the regulations according to its invention but over time and experiences, men have made a lot of contributions and the regulations have been much changed over time to arrive at the regulations we have today. Let's take more examples:

- Another example of Chrologic evolution: is the process from birth to death.
- Another example of Chrologic evolution: is the birth of galaxies until they disappear.
- Another example of Chrologic evolution: is the process of an election until the end of the mandate etc.

Take this simple banana example: The simple cut a banana contains a skin and the inner part.

Let's use the Chrology to understand by positioning ourselves on a tiny part of one of its parts we cannot detect all the shape and the limit of this banana.

This example is typically Chrology and being on several scales in the universe.

In our daily life, many examples of natural and artificial creatures allow us to understand these simple fundamental principles of limits throughout the universe describe by Chrology.

All truths we all believe are not true on other Chrological scales.

All presences or existence are influenced by several factors of other natural existences that we have difficulty mastering.

The existence of all these factors that influence all creatures is the Chrological presence of the external existentialist.

There is always something we call the engineer of any presence of existentialist who governs all our systems extra micro or extra macro etc.

>> *<<We can bring all the theories, but our big problem comes down to a mystery of Chrology, mean dimensional it will be very difficult for us human on our Chrologic part to know more beyond of our minimum or maximums dimensions.>>* Ulrich Ndilira Rotam

Consequence 1:

Any system of domination, or weakness on other systems, always has a Chrological time and will eventually lose its supremacy or its weaknesses. The consequences of these different facts, positive, negative, or neutral, synchronize to evolve and shift the whole to the original points in different very complicate designs in universe.

As evolution is Chrologic, all its different factors are also Chrologic in all these systems of domination or weakness.

Let's prove with an example of gravity at 10 miles from the earth or anything that uses the reactors to dominate the gravity to fly will turn long but will eventually come back if the reactors stop at the equilibrium point, which is the surface of the earth...

We have many examples in Scientifics version to describe this part with major calculations to show.

Another Chrologic event: You can lie, hide something to all in any way, in any dimension, or you can attempt to force all in the direction you want but you will be show out and lose all after the Chrologic time allow facts and natural aspect in different domain to be apparent.

This is very clear of the way all the Universe is built in different face: We see, experiment, live etc. the facts from lying even, from positive progress, from negative progress, or from neutral etc. in all universe available to us.

These are the realities men never know about because our domain is related to first Chrologic principles.

Consequence 2:

Any factor that causes negative or positive effects never ceases but instead continues to re-engender its effects in alternative ways depending on the Chrologic Time to continue to emit negative or positive effects.

Consequence 3:

Let's take a simple example of food. You put it in your mouth a quantity of food to eat, the activated teeth crush the food, and you swallow it.

- From the solid form you have gone from the crushing phase to the liquid phase.
- From there you swallow; inside the stomach several phenomena occur to digest the molecules.
- Then onto another phase you receive through your different tissues the food for your survival and another party to reject as waste.

In conclusion, we move from the solid phase to the liquid phase and the digestion phase of the molecules.

Chrologically these steps are simply explained by the existence of the **Chrologic worlds**.

The **transformations** are mechanical then followed by several positive or negative results.

It is also the example of a process of human birth, plants, animals, planets, galaxies, etc.

The **materialization** and **processes** are **generalized** in the universe for all **existences**, and all **presences**.

All that can exist can be conceived artificially in the universe.
Our brain evolves more closely according to our progress with our **solar system** in the **Cosmos**, as a function of time.

This is the reason why visual devices of the micro and macro world help us to understand certain visions of several environments throughout the **cosmos**. And the cosmos system lives and evolves in its entirety.

Throughout **Chrology**, the constants are at the base of the limits of the comprehension of our systems described according to the times and according to the evolution of the human and scientific progress called: The Limits of the **Leugs**.

Within the different principles of **Chrology**, one can build an artificial planet, an artificial microbe, an electron, a galaxy, and so on. However, this depends on our progress and our means of access.

All consequences that exist can be conceived as artificial in the universe and it is a logical law of nature that is described in the fundamental principles of Chrology.

It is necessary to know how to pose and interpret the results of the **equations Chrologically** to pass from one phase to another. And this process of calculations will allow us to create, innovate, and invent many new things. Some details will be given in the scientific edition of Chrology.

Chrology contains **six (6) constants** called the Human Constants on the **five (5) universal domains** distributed according to their limits in these five domains function of **LEUGS** which will be explained and applied in the **scientific edition** of Chrology with several calculations.

Chrology describes the information in all the **cosmology** that is shared in all five worlds, described by its first principle and the mathematical quantifiers of which are not exactly applicable in the five Chrological worlds. There is a single path in the five domains, of which explains all.

For an existence to be a **mathematical** meaning in Chrology it is necessary that the application of the factors of differences be applied and each door of study that we enter through the Chrology opens another door of reflection. These make the understanding, or the general view of the whole universe takes us very far to know many of its phenomena and many of its mysteries.

We can bring all theories to the universe, but our problem is limited to a Chrological, **dimensional mystery**, and these will be very difficult for us to know beyond our minimal or maximal dimension delimited by the system itself. Even for its original existence.

>   *<<We can never go further in the real quest for absolute*
>   *understanding of the universe because every human being*
>   *is an element of the thing and hold a small point as the*
>   *limit in this absolute to total comprehension.>>*Ulrich Ndilira Rotam

>   *<<Unfortunately, we live in a world where everything has*
>   *been artificialized to describe not only the true existentialist*
>   *principles but the domineering and material description.>>*Ulrich Ndilira Rotam

Any natural involvement will involve many other natural factors that if we all want to determine an inventory of Chrological events.

The inventory of **Chrologic events** is the determination of each event in the five parts described **Chrologically** according to its evaluations.

This counting is done in a complicated way because it is necessary to sum all different small parts in his five domains that bring from domain to domain.

Our presence in the universe is like millisecond or billion years compared to all its different absolute scales.

Our existence is based on a **K factors of existentialist** Chrological we are in a world admitted by the positioning of space and time in a K of existence to access the realities of all the universe in general and in particularity.

We cannot change what has happened Chrologically. The world we live in is an irreversible complex of functions of time, space, and K existentialist factors in the five **Chrologic world** described.

Through my research when I saw people who would minimize, discriminate, or believe themselves to be more important, by race, continents, countries, religions, social classes, money, completely upset me and push to a conclusion. If and only if they knew how much they are nothing with these factors of pride in all the cosmos and especially how it is represented in this complex system then they will never make these visions of differentiation.

Our presence in comparing throughout the universe is like a millisecond compared to all its **macros** scale and billions of years compared to its **extra micro** scale on all its absolute dimensions.

But our human pride distances us from the happier meaning of living in this colossal, complex wonder whose mystery of its dimensions are completely far from our door.

All existing information is preserved but the way to repeat it is difficult for us. We are tiny at the **Chrological scale** if we want to see so far, **macro and extra macro** we must set our limit to a foundation on our scale.

Can one read on time space and our **K existence** that the universe gives us to understand the whole secret of the universe? A very difficult question because Chrologically we will always be able to be limited.

The apprehension of the whole universe at a given moment is a complex and an indeterminacy, which is determined very globally as a function of time and space in the global world, which in turn is very difficult to determine in one of the domains described by **the first (1ˢᵗ) principle of Chrology** if we consider our classical methods.

The whole **universe composes** and **decomposes in a very complex** way that **leaves us to no chance to apprehend** it to **understand** it and **to study** it with **ease** in its **entire entirety at a given moment.**

Our limits of existences and understandings throughout the universe are limited to three areas described in the **first (1ˢᵗ) principle of Chrology.**

Chrologically consider methods, allows to consider all the intricate details and unnoticed details in the global world to apprehend and determine the universe in some instant gives if we are in one of the areas described by his first principle.

We are living beings and we are also the consequences of the transformations of the elements of the whole universe.

Chrology makes it possible to know our limits of understandings and existences because one can never exist or live on all the different existences of the entire universe.

> << *We can never hide forever any actions completed in this world. To get and see what happened we need to dig into the different element that existed at this moment to show us the true. Each part has Chrologic explanations and we need to assess and understand how to be interpreted to see the true.*>>
> Ulrich Ndilira Rotam

Chapter 3

Unique Model of all Searches Unified by Chrology throughout the Universe

Contents

The question of our diversity of knowledge and the question of total understanding of the whole universe

Human knowledge is an evolutionary **process** and very **difficult** to classify in a **unique setting despite** our attempts at classifications.

We have admitted several doctrines and facts with different conventions that allow us to separate them facing our knowledge's at different times of our existence and our evolution but within our universal convention there are several things that are interpreted and admitted with big errors and completely different from the absolute survey of the entire evolution of original universe.

We have made our knowledge into an absolute scale to explain the origin, the evolution, and the end of the universe, but we have missed many things that we do not understand and explain their important root cause.

These bring many questions to the forefront:

With our different doctrines, does everyone defend, in this domain, a force to explain the understanding of our existence within the universe as right? Not just on our scale of view and existence, but on the quantifiable scale of the whole universe.

Many of our **scientific** divisions and the studies of our existence for the absolute comprehension of the universe are still not convincing enough, especially if we let the time go, we will realize our huge mistakes in the reality that all the facts of cosmos show us. Even if we have understood and succeeded in explaining certain facts of the cosmos, our paths are still far to make us all confident about the facts, the existences, and presence in all the universe to explain with a single and simple method.

We are in a very **evolutionary** and very **complicated system**. To explain in a single method that brought together all our knowledge we are brought back to a fundamental equation.

>   *<<Do we really know with all our possession the deep understanding of our existence in this complex system?>>* Ulrich Ndilira Rotam

To unify all our **different disciplines** and **sciences** in a **single model** to allow us with a **unique formula** and to explain the whole universe as a whole, **Chrology** has introduced a quantifying **factor** called **LEUGS.** This is the set of all our progress, efforts, discoveries, inventions, interpretations, progressions of knowledge throughout the universe, formulas in our world, in the **tiny** world, in the **macroscopic** world, in the **extra-microscopic** world, in the **extra-macroscopic** world of the known cosmos, existing in the **first principle of Chrology.**

All the discoveries, the inventions, the products of the researches on all the domains known by the man classify in Leugs in all the **Chrology** and all those who are known of the man is absolutely on a part of the first principle of Chrology and the **Leugs** evolve with time in the **Chrologic worlds.**

Consequences: Mankind will know more about the basis and functioning of the **Cosmos** according to the evolution of the **Leugs** in each field distributed according in the **first (1ˢᵗ) principle of Chrology.**

Chrology describes all the existences and evolutions of the universe in its complex diversities, and all those we study, or have been studied, can be found in one part of his division. That is why it constitutes as a unique model of the whole universe for which all our knowledge and studies are found.

In the **entire universe** the place where man lives are homogeneous and favorable to his survival, we must not live in certain domains which are described by the first principle of **Chrology**.

The **universe is a complex natural system** that gives us more access in the **Normalogy** to understand the different **synchronizations** and the different **complexities** of the whole existence.

All is the sum of **existence** and the **presence** of something in the **whole universe**.

The **transformations** of energy or the transformations of matter are only the simple **principles**, mechanisms, and transformations that evolve towards a **positive**, **negative,** or **constant** sense and are developed by a complex natural law in **Chrology** we will develop in **scientific publishing**.

Chrology Picture 4

The **Micro World** in its **domain Chrologic Principles** are same in this **micro world**.

Matter accumulates to form something, and this thing is only an **existence** and this existence must be **present** somewhere in all the **greatness** of the **universe**, so the big questions we ask ourselves are:

- How is this interpreted in a single model?
- How are they formed in relation to the various feasibilities of the things in a natural and original way?

It is the **fundamental question** that we want the answers to elucidate before certain **mysteries** of the **whole universe** in its different **forms of existence** and **Chrology** will help us a lot with its principles to put the lights on his understanding and clear them up. **Existence and presence** of the **Micro world** in its domain.

In all these worlds the **Chrologic Principles** are same in this **micro** presence it's in this way the **original existence** allows us to understand and confuse us on **existences of everything**.

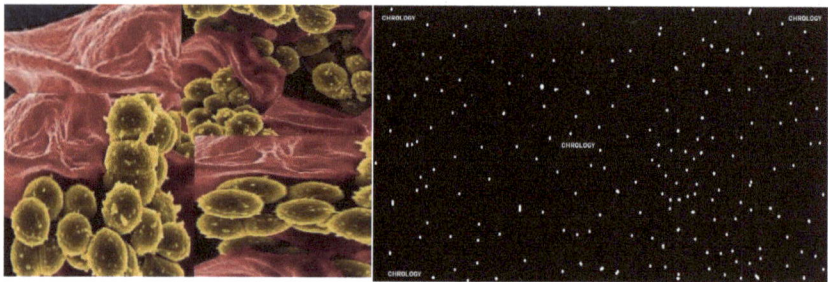

Chrology Picture 5

In this portion of **micro existence** all **Chrologic law** are present we can always assume five existence of world in this sample. This is one small example on infinity existence.

1) **Chrology** describes a person as existence and presence, a house as existence and presence, a plant as existence and presence, a planet as existence and presence, a galaxy as existence and presence, a cosmic foam as existence and presence. But the way in which nature or cosmology presents them, which differs and classifies in the domains described in the **first (1ˢᵗ) principle of Chrology.**

2) **Chrology** will tackle these questions:
   - **Why is there something present in Michrology?**
   - **Why the other exist in Machrology?**

So that Chrology itself exists and all of **Cosmology** can exist in its original state to be understand.

3) **Chrology** provides a **vision**, a scientific description, literary description, artistic description of Universe.

The dark energy and energy known in our standard knowledge exists in Chrology to understand the mystery of the Universe, but to explain in another way in the **five (5) parts Chrologic.**

4) When an element or elements react with one another, or others, in any way on any area in the cosmos, the results will give the sum of their existences and presences, then their energies and their materials of the tiny scale to the macro scale in the **five (5) Chrologic** worlds and it is in this unique way that the cosmos shows its existence and presence.

All human knowledge is incorporated in Chrology

All the laws rigorously conceived on all the domains of the human researches with the scientific logic, literary are all incorporated in a limited way and distributed throughout the Chrology.

This is the example of Hubble's expansion law that takes us to the understanding of our Normalogical world by moving away to present the other celestial objects in a space-time of the Machrologic world.

- Another example of the fractal overview of the different entities and packages of quanta, which are all incorporated in the Chrology.
- The law of gravitational attraction of Newton is present, throughout the Chrology but in a different way as redefined, by the general relativity in the same Chrologic field.
- The constants: cosmological, gravitational, Planck, Boltzmann, Rydberg, light are all divided, in the different fields of Chrology and explain dif-

ferently the realities of the universe, in a typically Chrological model, if we envisage a real study in a simple way, to go further in the comprehension of the global world.

- **Chrology** being a model which explains the **domains** of **existences** and **presences** of all, overall of the universe, all our known disciplines, are all distributed in the five (5) world or domain of the **first (1st) principle of the Chrology**. It is an extraordinary and very complete understanding in very simple way to decorticate how the **whole existences** can be and present to us.

Chapter 4

Chrology is the Science of All Sciences

Content

1) What is Chrology?
2) Chrology describes the whole Universe by a Law of Evolution.
3) Chrology describes the divisions of the Structures and the Edifice of the all Universe.

What is Chrology?

The simple definition of **Chrology** is the **theory** of **unification** of **all knowledges**, all **discoveries**, **all great laws**, all **sciences** on the **whole universe** in a **single concept** showing the **different faces** of the **whole system** in all its **dimensions** according to our **evolution**.

Chrology in his theory helps us to **know**, to **classify**, to study, and to show all **existences** in its different particularities and to meet the different laws, **the different diversities** that the universe system possess so that the man can accesses with his **knowledge**, **intelligences**, his **genius**, **Universe** materials without however distorted the original concepts of the interpretations of the system in all his **diversities**.

Physic Chrologic is a **concept** with its **principles** showing the **different doors of horizons**, environments and the **understanding** of the **constituents of any system** of the **universe** in its **originality** and its **set**.

The five domains described by its **first (1st) principle** exist in **infinitely** large numbers, **infinitely** small and fixed in all its set so that this **physic Chrologic** can exist in all its **originality** to explain without modifying all **things** in all **universe** in its **original concepts**.

Chrology also opens the **knowledge of the universe** to Man with all his different constituents that can be reproduced or transformed artificially. Another term called **Leugs** is introduced in this **new concept of Chorology**.

All existences come from substra which is the composition of all existence in the whole system of complete universe. Men comes from the **substra,** which is the composition of one element present in the universe.

We are limited in our world of **logic** that can be applied in all space and time that is for us but very difficult to come to understand this wide world that is beyond our space and time, or this thin world that are outside of our space and time.

If we apply **Chrology** one will understand on a very large and on a very thin peculiarity of the **whole universe**.

Let's look from our position to the sky we see without the binoculars or artificial satellites that a limit must exist, but if we take the devices to look at this limit, we realize that our limit was not a limit but that there is still a wide field of view to another limit.

With the help of our spacecraft and satellites, we look at this limit we go even further but that the limit we thought with these devices was not a limit but that it opens yet other fields of vision.

If we still had another technology to keep looking at this limit, then it would not be a limit we thought but there is still something about our wide field of vision and so on.

And if we turn all these mink towards the tiny fields, there will still be the same situations and we will evolve from one vision instrument to another.

Consequences we **exist** and **live** in a **Universe** very **large** and **infinitely** infinite to **our scales** that the same nature has generated and granted us to gain access. But **Chrology** allows us to understand the different limits facing our different efforts of understanding and our different **quests' answers**.

The **LEUGS** introduced by **Chrology** is all our scientific studies, our vision efforts, our interpretive efforts, our knowledge, our intellectual conventions in the face of a presence or a **physical** situation of any **natural** and **artificial** material in the **world**.

Any part of the universe or many things happen and manifest that we do not have direct access to them to contribute to the evolution of the whole system.

The Leugs that man can conceive, or reproduce on the natural constitute to the artificial for the evolution of the whole system brings about a big absolute difference in all the system of universes:

The artificial is different from the original
and the original is different from the artificial.

And those lead her to non-equivalence throughout the universe:

ARTIFICIAL ≠ ORIGINAL

<u>AR</u> Different <u>OR</u>

Natural Chrology≠ Artificial Chrology
=> Co ≠ Ca

Artificial Chrology ≠ Natural Chrology
=> Ca ≠ Co

Here is an example of two completely different cases to illustrate the Chrology:

- Let's take an asteroid that evolves in its world with all its parameters of study is one.
- Let's take a microbe that lives and evolves in its world with all its parameters of studies.

Overall studies are done in a way where we use our exact sciences in the Macro world and the exact sciences in the Micro world to explain them to our conventions of study.

But if we want to go deeper into more in-depth details, we need to make use of Chrology to fully understand their two (2) worlds with parameters such as their different energies, their different movements, their different presences, and so on.

The manifestations and evolutions in each system, as described by the first principle of Chrology, contribute to the evolution of the entire system of Cosmos in a bigger and minuscule scale.

Finally, the entire universe is only Chrologic and one evolves from Chrology to Chrology.

Consequence:

There is the existence of **two worlds** which give the secrets of understandings of the whole universe in different ways the existences of any presence on all its dimensions are present in these two worlds:

- Global world
- Chrologic world

The existence of these two worlds causes a difference that is never equal in reality:

GLOBAL WORLD ≠ CHROLOGIC WORLD

Chrology describes the whole universe by a law of evolution

The whole universe is governed by a law of evolution. Time and space contain and accompany every system in the universe towards its evolution.

From the smallest to the most gigantic system in perpetual evolution, no one is a complete system that shows all the secrets of the universe.

Example: Let's take a banana. It contains a skin and the inner part.

If we shrink ourselves down to be at a very tiny size and position ourselves on the tiny part of this banana from an inner zone to the atomic size, we fail to detect all the shape and the bound of this banana, however the banana is a finished system but very difficult to know if we are part of this banana.

This consequence is applicable on the big part if we position ourselves on the external skin it will be difficult to understand the tiny parts and to know the total form of this banana. This example is typically Chrology.

Chrology is everywhere in the universe. It is in a gravitational, electric, magnetic, electromagnetic field, in a wave, in our body, in water, in fire, in fluids, in the basement, in space, in a smoke, in the cosmic foam, in the atoms, in the electrons.

All material all existence in the cosmos is a Chrology. The Chrologic calculations are done if and only if one arrives to grant him a mathematical physical logic with the formulas. Then one can calculate his Chrology into different Chrology: Those generate infinite calculations of Chrology in the universe.

Any system in the cosmos is in the five (5) parts described by Chrology and are naturally governed by their natural laws that can be described by physics without modifies mathematically the original concepts.

All material in the cosmos exists in several dimensions and the order of magnitudes is governed by Chrology law.

Atoms are macroscopic in another dimension and microscopic in another dimension, infinitely large in another dimension and infinitely small in another dimension.

Time and space are everywhere throughout Chrology but irreversibly in all five (5) configurations described by the first principle of Chrology.

We are in a very limited part of the universe, like microbes, electrons, are also in their very limited worlds. The mystery is announced in a Chrological way: from gigantic to minimal or from minimal to gigantic.

The different domains in different parts of the universe have different formulas and his studies described by the man called LEUGS are very broad with contradictory principles. Let us study by comparing the different domains the realities of nature become a Chrologic complex.

All theories show the different attempts of the beginning of the universe, but we are only confronted with the complication of our dimensional limit.

A microbe that tries to understand the whole system of the cosmos will carry one or more theories of its big bang with a periodic table of completely different elements of our which explains little of the universe compares to its dimensions but if its studies are done by Chrology he will be more advanced in his search for understandings.

For example, a fish that lives all its life in the water or bacteria in his world will never understand if the universe is expanding and our large space exists even if it emits the best technologies.

We all-natural creatures evolve in our worlds we are part of many parts of something that evolves in a Chrological way. Space is everywhere and is no longer the same.

Each subject has a reason to be, but it is this reason that is difficult for us to determine.

These new conceptual approaches are real interpretations that allow us to go even further with new points to reflect more and more broadly and deeply on a division and vision of universe contains study or I like to say our system ... those are simpler and more complex studies.

It opens a new direction to meet all the systems of the cosmos in a context more appropriate to man on the way where all the existing ones can appear and be interpreted with our intelligence, our comprehension and with our scientific vision to understand and explain in his entire original environment.

It will also allow the opening of several discoveries and explanations of the system of the cosmos which are already validated, or which will be very broad and accessible in the different divisions. But many of the disciplines and research that have already been realized in much of this theory will be distributed in the different Chrologic subdivisions.

Nowadays we have a very large number of sciences which allow us to see the complicated and easy things, in this new conception of Divisions and subdivision, the sciences and the literatures conceived are distributed just in three (3) parts that we have in the first principle of Chrology.

Positions in the universe influence much of the understanding of Chrology which is different from big bang theories, relativity, and so on. A big explanation on this part will come in the part of Scientific Edition.

Technologies are factors that change over time. In the past, the world existed very well with its technologies and the people were proud.

The **Leugs** are the determinations of all those for whom Chrology explains the system.

Building a device that allows you to determine the different factors Chrologic is the most important thing to advance in our research quest throughout the universe and opens access to the answers of several questions and to the understanding of several mysteries of the universe.

If we know the Chrological Factor in a field we are studying, we can determine the initial conception of what we plan to study.

This determination can contribute to a great advance on various research inquiries in further understanding of the universe.

Let's take a country where the laws allow governing a nation. All these laws are designed as Leugs to determine the existence of this nation according to time and these Leugs or law will also change according to the time that is why all those who must maltreat this so-called beneficiary on any convention are all losers because the consequences extend in different Chrology and as they do not understand they will pay according to the time.

All our sufferings and bad evolutions, in the negative sense, are the consequences of some poorly done things.

All our successes, our pride, and so on are the consequences of a few well-designed things.

This Chrologic reasoning applies to all organizations, continental, planetary, and different structures designed.

> <<*I am more interested in the finest or greatest particularity that others are not interested in that can have the same results in any organization to determine the best way and the best functioning.*>> Ulrich Ndilira Rotam.

This differentiation leads us to always be more successful comparing to most things we have achieved in our lives than with little or more means.

The chrology describes the divisions of the structures and the edifice of the universe.
The structures and the edifice of the composition of the universe on all its generality are not a small and simple complex of organized order.

The universe is composed of several parts whose studies of its building differs from one environment to another, from one horizon to another horizon.

In the world of macro, many sciences have emerged, studies are made with fewer errors to our understanding scales that why the macro world is spreading with much more numerous and affordable sciences that allow us to understand certain mystery of the world.

The cosmic system is mostly a horizon system until we no longer arrive with our scientific progress to continue to measure this horizon if we position ourselves at one point on the earth or somewhere in his any existence.

Several theories have emerged in this area to try to clarify the absolute understanding of everything. But the entire universe hides his deep secrets and we stayed with several problems of methodologies and technics to centralize his mysteries from his big views to his tiny existences see universe picture bellow.

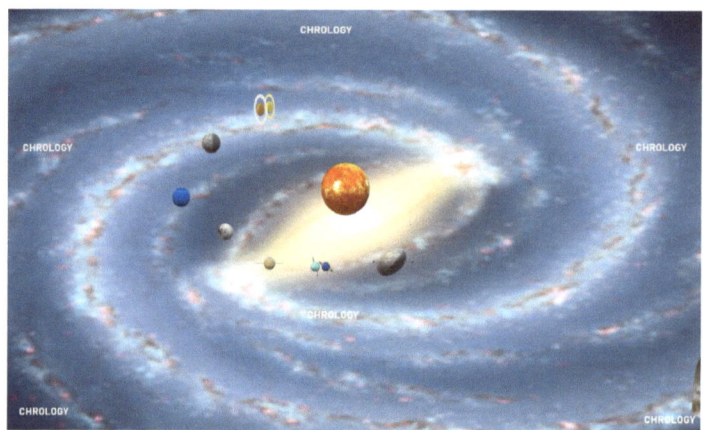

Chrology Picture 6

Let's use our precious sense organ, vision, takes us to see up to certain limit of space until is no longer allow us to see more and to no longer understand the concept of all universe to interpret things scientifically.

If we switch using our vision instruments, we find ourselves again in the same situation.

Reason why the methods propose by Chrology defined all the set of studies in universe then to position all our studies, our sciences and our existing knowledge, in their descriptive mode of existence.

Chrology is the science of all the sciences described in his theory with the five domains of studies that are Chrologically performed throughout the universe and describe it originally:

- Substranormalogy,
- Substramicrology,
- Substramacrology,
- Substra-extramicrology,
- Substra-extramacrology.
- Substramicrology

In the microscopic field cut into micro Substra, we define Substramicrology as the group of all the sciences that study the natural or artificial creatures that are in the micro Substra field.

Substramacrology

In the macroscopic field divided into macro substra, we define Substramacrology as the group of all sciences that studies in all sense all natural and artificial existences that are groups in this macro Substra field.

Example:

Astrology, Aeronautics, Astronomy.

Substranormalogy

In the normal field as we access in our positions cut into Substra Norma we can define the Substranormalogy as the group of all the sciences or literatures that study in all the direction the movements, the energy, the forces, the positions, the different behaviors natural and artificial creatures that are in the Norma Substra.

Example:

Kinematics, Geometry, Arithmetic, Architecture, Sociology, Geography, History, Topography.

Substraextramicrology

In the Extra microscopic field divided into Substra extra micro we can define the Substraextramacrology as the group of all sciences or literatures that study all the limit of natural creatures in Substra-extramicro. These are being natural creatures (Existences) invisible by our electronic devices of vision or optical microscopes which tend towards the infinitely small.

Substraextramacrology

In the Extra-macroscopic field divided into substraextramacro, we define Substraextramacrology as the group of all the sciences which studies in all the direction the natural creatures exaggeratedly big or small, but which are invisible to our glasses, binoculars, or all the instruments that allow us to see very far in the universe that we now have.

Chapter 5

Fundamental Principle of Absolute Existence of the Whole Universe. First Principle of Chrology.

Content

The First Principle of Chrology

The **first (1ˢᵗ) principle of Chrology,** also known as the Fundamental **Principle of Existence** of <u>all</u> is composed of **3 postulates** which tell us:

Postulate 1

The **whole universe** exists and decomposes **Chrologically** to **appear** to us in **five**

largely world called **Chrological worlds (CW)**^{MC}.

- Normal World= $^{MC}G_0$NW

- Microscopic World = $^{MC}M_I$MiW

- Macroscopic world = $^{MC}M_\bullet$MaW

- ^{ME}m Extra-Microscopic World = $^{ME}m_e$eMiW

- $^{ME}m_\bullet$ Extra-Macroscopic World = $^{ME}m_\bullet$ eMaW

  CW = NW + MiW + MaW + eMiW + eMaW

Postulate 2

All **creatures**, all **existences**, all **presences**, all **particles**, all **movements**, all **energies**, all **forces**, all **manifestations**, all **lives**, all **studies exist** and **present** themselves in these **five 5 worlds** throughout the **universe** and **evolve Chrologically** as a function of **time** and **space** in the face of its **global presence**.

The **Chrology** of an element ϕ in the entire perceived **universe** is the **sum** of its **existence** and its **presence** in one of the domains described by the **first postulate**:

$$C\phi = E\phi + P\phi$$

With

$C\phi$ Chrology **of** ϕ

$E\phi$ Existence **of** ϕ

$P\phi$ Presence **of** ϕ

Postulate 3

The **five (5) Chrologic worlds** of the entire **universe** described in the **first postu-**

late are irreversibly changing in their size, leading to **K existence**, **time**, and **space** to accompany them in different **ways**.

Each part constitutes a unique model different from the others in its existence and its limits resulting in an absolute inequality so that the whole system exists in its structure and together:

Global World ≠ Microscopic World ≠ Macroscopic World ≠ Extra-Microscopic World ≠ Extra-Macroscopic World

GW ≠ MiW ≠ MaW ≠ eMiW ≠ eMaW

$$\mathcal{M} G_0 \, \mathcal{M} E_{m_2}$$

Consequence 1:

To make a **Chrologic study** in the universe we must determine and fix the absolute origin of the **Chrologic** marker these leading to a study of **Chrology in Chorology** to allow us to know the size in which **domain** we are doing the study.

These **consequences** will allow us to develop and explain several **mysteries of the universe**.

The **five Chrologic worlds** in the first postulate are described in a unique model to explain all **cosmology** to exist in its originality:

Existence of Different Worlds

The glimpse of all **existences** and **presences** throughout the universe is an **absolute** to speak **all the universe**.

The different **births, deaths, destruction, evolutions, transformations, etc.** is in these **five (5) worlds** according to **Chrology**.

The **five (5) worlds** are:

Global World

The Global World in **Chrology** is the subdivision part called **Substranorma** that explains the existence and evolution of all who are in this **normal domain** as existence with its natural and artificial laws.

Macroscopic World

This world in **Chrology** is the divisional part called **Substramacro** that explain the existence and evolution of all those who are in this **macroscopic domain** as existence with its natural and artificial laws.

Microscopic World

This world in **Chrology** is the subdivision part called **Substramicro** that explain the existence and evolution of all who are in this **microscopic** field as existence with its natural and artificial laws.

Extra Microscopic World

This world in **Chrology** is divided into parts called **Substra-extra micro** that explain the existence and evolution of all those who are in this **extra-microscopic** field as existence within its natural and artificial laws.

Extramacroscopic World

This world in **Chrology** is the divisional part called **Substra-extra macro** that explain the existence and evolution of all those who are in this **extra macroscopic** domain as existence with its natural and artificial laws.

The origin of the divisions can vary and always **Chrologically** so that everything exists in an original way without modifying them.

Consequence 2:

On earth to make **Chrologic studies** in the **universe**, these divisions are adaptable to its original model to allow us on our planet earth: if we want to see how we with our thoughts are, our laws, our logic, our formulas ... And especially what the universe allows us to access to the limit of our efforts throughout the system.

Several **calculations** will be developed in the **scientific publishing** parts on the different **Chrologic calculations**.

Chrology in the whole universe shows nature with irreversible visions of the position that one finds oneself in all of it. These are non-relative studies.

What an existence looks on the planet earth is not the same thing or the same conditions as if the other existences on other planets, galaxy, micro world look at or conceive: The Leugs are always irreversible.

Experiments:

Let's do three experiments to better understand all these consequences and concepts:

Experiment 1

Let us take **eight (8) men** in our earth they have same age with the same diplomas in different disciplines of different intellectual aspects with the same **devices** or the **same sense** of **technology** that they can use in the different environments and we assume to send them for study of understanding the different parts of Universe. Let's send these men **two (2)** by **two (2)**:

    a)   Send two (2) men on **different planets**.

    b)   Send two (2) men to **different galaxies**.

    c)   Send two (2) men to **different tiny worlds**,

    d)   Send two (2) men to the **top of the Eiffel Tower**.

The consequences of the results of the vision samples and different results of our **eight (8)** visitors in these **five (5)** world will be completely different globally but **Chrologically** the same:

Note 1:

The **first (1ˢᵗ) principle of Chrology,** remains the same for all eight **(8) visitors**. Each traveler will always have to conceive and recognize his world in **five (5) parts** as in the **first (1ˢᵗ) postulate** of Chrology. Every visitor will continue to see his world always defined in Leugs.

Note 2:

The results according to the tendencies of approaches will be completely different the dimensions, the spaces, the movements, the energies and the Leugs designed will not be the same according to the different trips of our visitors.

Note 3:

Approaches to understanding the cosmos of their worlds will all be complicated for all 8 visitors. It will always be difficult for all visitors to understand exactly the systems of the Cosmos and everyone will always be limited.

This is the way all universe is conceive and appear to us in any circumstances, with any technology we dispose.

Experiment 2:

Let's position ourselves on the Eifel Tower or Statue of Liberty and make several circles of visions samples in all directions.

Note 1:

None of the samples of visions will be the same and irreversible visions will always be understood according to the time elapsed.

Note 2:

We will always arrive at an interpretation for each sample of visions. But the understanding of the entire cosmos will escape us.

To facilitate the understandings of **Chrology** through the entire universe to my readers, there is a unique field of understandings on the distribution of our different disciplines in an original order to our existence and original in the entire universe without modifying it.

In **Chrology** the approaches of the view of the cosmos and its components are visible when the light wave packets reach it to us in relation to our brain and we talk about the existence of a few things: Group of the visible, invisible when the packets of light waves do not reach it to our eyes to transmit to the brain and we talk about the **nonexistence** of these things: Group of invisible.

But there is a problem with the look because if our eyes do not see the atoms directly, without a microscope or look at a planet that is very far away at millions of light years, without astronomical glasses or satellites then the man concludes that there is nothing like existence and he sees nothing.

The **NOTHING** possess an explanation well defined in the group of invisible according to the explanations of **Chrology** those implies a physical existence and non-physical existence.

Experiment 3:

Let us take a liberated energy that manifest itself in the normal field of vision of the man (**Substranormalogy**) and provided with a heavy consequence of materials destruction.

These **complexes process**, and liberation of massive energy are explained in the theory of Chrolos in several ways, in successive ways considering all Chrologic parameters to make its explanations very easy.

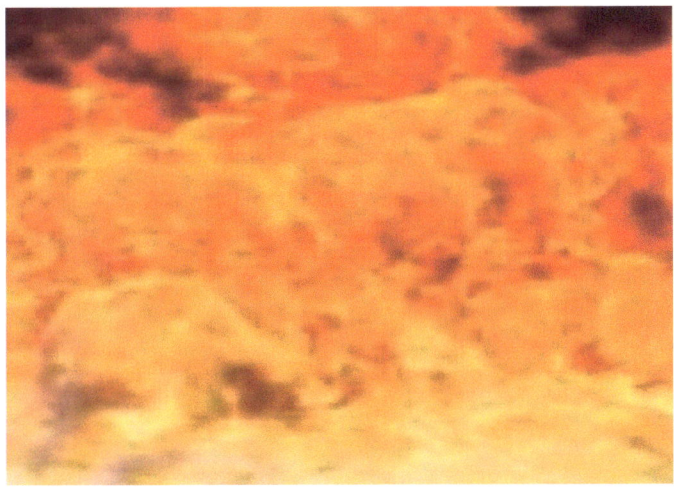

Chrology Picture 6

Note 1:

This energy manifests itself differently in each area described by Chrology, there will be several processes before we look at the explosion.

- In the tiny world atoms collide.
- On our normal position we recapitulate a force and exorbitant heat liberation of energy that destroys everything in its path.
- In the air the effects of dust and fumes navigate and disappear in part of the **Chrologic domains**.

Consequence 3:

In this example, the manifestations are not all the same. In the tiny space, the colliding electrons are clean in this area doing their job, but when we arrive at the normal position we receive energy liberated with great force this is another different manifestation than the first, in the third domain in the space the dust and fumes do not give the same effects as in the two previous environments always a new form of manifestation happen. And the consequences are all different.

To reconstruct all these stages, the **theory of Chrolos** helps us to understand these facts very easily.

Important Note:

Each time we change configuration in this world corresponds to a specific change of domain according to each Leugs to study with **Chrology**.

It means to pass from one study to another of all-natural creatures it is necessary to understand that the system of the studies must first be determined in what parts of Chrolos? Then the Leugs and the studies in question can be easy determine.

Movements, Energies, Presence are accessible and different in each area in our existing universe.

Chapter 6

Fundamental Principles of Unification of All Studies in Universe.

Content

Introduction

Approaches to the whole universe with all its tiny systems, macros, and its natural constituents on the earth globe, on the different planetary globes, galactic in its generality are real on all its systems of existence from environments to other environments and from horizons to another horizon.

They present diversified beauties, very pretty and very complicated to contemplate that make it difficult to study with more precision in a very limited field and in a very specific field like our position on the planet earth when we envisage a scientific study to understand everything together.

This is the reason why we need new ways to apprehend, approach, and new formulations to make his studies simple and understandable to better interpret them, understand them with all our different means, formulas to explain them in unique and simple scientific concepts in a very sophisticated way in the face of all our discoveries, our progress, and limits of access to master it in our deep quests of absolute understandings.

Scientific studies, literary conceptual, and theoretical for the determinations and the comprehensions of all natural existences, the relativizations, the quantifications, the gravitations, the different chemical and biological transformations, the classifications of the studies of a real position in a given time, the Absolute studies of the constitutive elements of the cosmos, the big bang theory, the theory of inflation, the theory of the strings are all in this new and unique overall model of all the sciences of our whole history and human existence described by **Chrology** and is verified everywhere on the whole universe.

The applications of all our **laws, properties, principles, axioms, formulas, conventions, in physics, chemistry, biology, mathematic, literary,** in all his compositions are found in various ways and classified in the **different conceptions of Chrological Science**.

The different orders in this unique **Science** are linked in a simple way by obtaining complex or complex results by obtaining simple results.

All the applications of a human knowing describing any portion of the **universe** in mathematics, physics, chemical, biological, literary, artistic is a part in **CHROLOGY** which is elaborated in a logic much widened and deepened in relation to the realities that the universe presents us as to access.

Also, the way or the whole system evolves and presents us its degrees so that we can control and be able to understand them, to interpret ones, to transform artificially and to maniple or to uses as a function of the time for any evolution.

The **Chrological Principles** are at the **base** to **enable** us to **understand** and **determine visions** of all those around us and the dimensions that we do not have access to.

CHROLOGIC physics in all its presentational and immeasurable beauty of the whole universe shows several mysteries of existences and presences of the different elements of the universe.

Second Principle of Chrology

The **second (2ⁿᵈ) principle of Chrology** called the model of **unification of all our knowledge**, or called the **unification of all the sciences** called **SCIENCES CHROLOGIC** tells us:

All **universal studies** admitted with its **different varieties** and **complexity by man** throughout **the universe** are **presented and classified** in **the five domains** described in the **first (1ˢᵗ) principle of Chrologic**.

There are **five (5)** elements that constitute the existence of all universe to any study we called it **CHROLOGIC SCIENCES** which are:

1. Substranormalogie,
2. Substramacrologie,
3. Substraextramacrologie,
4. Substramicrologie,
5. Substraextramicrologie.

We will come back to clarify with several demonstrations on our scientific edition to explain, disassemble, and to get out the list of all our different areas of knowledge's admitted on each parts **SCIENCE** by science.

The five **CHROLOGIQUE SCIENCES** explains and unifies all our **human knowledge's** in a **single domain** to understand the **mysteries**.

> *<<All existences are only a mathematical quantifier known and unknown to our dimension that we measure or present ourselves intuitively in front of our intelligences, our devices, our knowledge, our intellectual experiences, and especially our existence to be able to apprehend them, to understand them, and to live.>>* Ulrich Ndilira Rotam.

In this principle the fundamental basis and the multidisciplinary approaches of the wider vision of the Cosmos that are summed up in each part of the Chrology.

We can never **understand** a manifestation, a presence, or a physical, biological existence, material on the scale of universes without however determining in which Chrologic discipline we can study it.

That is why **many things escape** us in our thoughts and profoundly with our progress because the determination of its classification in **first (1st) principle of Chrolos** is not really carried out on a **definite scale**. And one seeks to understand the **principles** in a **field** that is not precise in the **Chorology** called **GLOBAL DOMAIN**.

This part holds the key to **explaining many mysteries** of our world our existence and even what is hidden.

The Third Principle of Chrology

The **third (3rd) principle of Chrology** also called **principle of inequality, principle of different world,** or **principle of confusion of presences and existence** in all the **universe** tell us:

The integral applications of our knowledge's, our experiments, our scientific developments, **the laws of science** are no longer the same in the **five (5)** scientific distributions field described in the **second (2nd) principle of Chrology**.

Whenever we study an **existence** in the **different parts** of the **first postulate** in the **first (1st) principle of Chrologie,** we will always tend to see it confused in the same **first (1st) principle of Chrolos.**

Everything is just a **Chrology** and we go from **Chrolos** to **Chrolos.**

Impacts

There will necessarily be the **existence** of **two (2) worlds** that are **absolute** and **different** so that the whole universe **exists** in its **original geometry:**

- The Global World
- The Chrologic World

$$GW \neq CW$$
$$CW = NW + MiW + MaW + eMiW + eMaW => GW \neq NW + MiW + MaW + eMiW + eMaW$$
$$=> GW \neq NW \neq MiW \neq MaW \neq eMiW \neq eMaW$$

The **Global World** gives us an **Existentialist** understanding, and the **Chrologic World** gives us an understanding of **horizon** to **horizon**.

- The scientific conceptions in these two worlds no longer have the same dimensions: Each part has its dimensions delimited with its laws so that the whole system exists and contribute to its absolute evolution.
- Time and global space are no longer the same as time and the Chrologic space, everything happens in a different way with several factors of studies according to their existences throughout the whole universe so that we can reaches his absolute understanding.
- The laws of physics must no longer be the same they must present concretely and differently in the Global world, and in the Chrologic world it's one of the consequences of the **second (2nd) principle of Chrology**.

These two worlds throughout the Cosmos allow us to perceive the meanings of the different structures of the entire universe with the scientific laws designed to explain them without modification.

In the global world we always continue to build new **LEUGS** to approach and understand the scientific studies in each part of the **first (1st) principle of Chrolos**.

Therefore, we are limited to discover all the complete mysteries of the Universe in all its immense compositions and diversities.

Global World:

To understand with a simple explanation of the global world, we must refer to life with all its scientific understandings that we presently access now as well as the limits that we no longer perceive with our present standard means.

It is the comprehension of the life, with our present reach with our organs of sense, vision, and our eyes, unaided by any device of visual correction whatever. We all want to see or understand the system of the complete cosmos.

The **global world** is summed up finally to a normal world very simple to all the presences of the creatures in the normal field or we fix like study for the accesses on any field of the universe.

All who exist leads us to better understand and apply with laws, relations, axioms, theorems, properties to study them.

With the global world, it is very easy to understand a natural phenomenon and interpret it with laws or formulas already presented to us however; it is very difficult to understand the complexities of nature and interpreted with the laws or formulas.

Example:

Earth, moon, sun, sea, forest, mountains, skyscrapers, tunnel, derrick, crowd, fight, race, football match, Rugby, an insect, an elephant, a fish, a flash of light, a field, a cloud, a tide, a rainbow, a coffee green, a computer, a smoke, dew, a text, a sign, a symbol.

Chrologic World:

To fully understand the **Chrological World (CW)**, we must refer to the **first (1ˢᵗ) principle of Chrolos**.

The **Chrologic World (CW)** is very complex and difficult to understand and access but to make it very simple here is an explanation: In the **Global World (GW)** we can always see and understand everything we look at and have access to interpreted and to make his studies more scientific or literary with the laws the formulas and our properties. It is the study of the normal field or the normal world.

When one leaves this normal or **Global World (GW)**, one directly reaches the **Chrologic Worlds (CW)**.

- All those found in the **microscopic world** such as time, space, movements, forces, energies grouped in **Substramicro** can no longer be explained in the global world. They are part of the Chrologic world in the **first (1st) principle of Chrolos**.

- All those found in the **extra-microscopic world** such as time, space, movement, forces, energies **Substra-extra micro** group can no longer be explained in the global world. They are part of the **Chrologic world** in the **first (1st) principle of Chrolos**.
- All those found in the **macroscopic world** such as time, space, movements, forces, energies, materials group in **Substramacro** can no longer be explained in the global world. They are part of the Chrologic world in the **first (1st) principle of Chrolos**.
- All those found in the **extra-macroscopic world** such **as time, space, movements, forces, energies, materials** group in **Substra-extra** macro can no longer be explained in the **global world**. They are part of the **Chrologic world** in the **first (1st) principle of Chrolos**.

Intuitively we will always have the **Chrologic** and **global world** everywhere throughout the universe, but their determinations depend on the **reference** of the **Chrologic studies**.

Fourth Principle of Chrology

All **quantifications**, all **spins**, all **determinations**, all **studies** of an **existence** in the **universe** described in the **Chrologic world** are no **longer** represented in the **same** way.

In all **five (5) parts** of the **Chrology** they are in **multidimensional** ways which **space**, different **manifestations**, **movements**, **energies**, **forces** are all different so that each **Chrologic division** has his **time**, **his space**, **his movements**, **his forces**, **his energies**, **his life** for all the **universe** system to exists **originally**.

Therefore, we cannot easily understand deeply the **system** of **Cosmos** at a determined position with **precision**.

A simple study in a **single field** of **Chrology** informs us and gives a limited understanding of the **whole universe**.

The strength of an electron is different from the strength of a planet.

Example 1

The **time** and **space** occupied by the **movement** of a quark, an electron, a proton, a planet, a galaxy, a cluster of galaxies, a cosmic foam are never the same.

This example is applicable also if we compare it to a person who is on the top of the Eifel Tower, on the top of the freedom status all moving as well then all will be different in the universe.

Fifth Principle of Chrology

In the **universe** there exists a factor of **determinations** called the **'Chrologic Factor'**, which makes it possible to know the **quantifications** of a **presence** of an **existence** compared to other **existences** or **presences** in the **universe**.

Let **A** and **B** two existences in the universe. The **Chrology** of **A** is different from the **Chrology** of **B**.

$$CA \neq CB => CB \neq CA$$

For **A** to be equal to **B**, we need the **F** factor, which is the original aspect that qualifies **A** and **B** in the universe to compare to their two **different worlds** for their **stabilities** in these two parts of the universe.

FC Chrologic factor of universe

$$CA \neq CB => CB \neq CA$$

$$CA = CB => CA = FC \times CB$$

FC Chrologic factor of universe

CA different from **CB** always in Chrologic domain of study.

Consequences:

The **Five (5ᵗʰ) Chrologic Principles** are the **fundamental** and **representative bases** of **any system** that **exists** on **any part** of the **universe** and **constitute** the great synchronization of a **gigantic** or **tiny** access of the **cosmos** for a great understanding of its **existence** and **presence** in all his complete system.

There will not be humans like humans on earth on other systems. But there will be humans on another system that be represents by another way if to compare to humans on the planet earth.

Example:

For an **asteroid** to be able to travel with great force from distant galaxies to manifest itself on the surface of our planet earth Reference Statute of freedom.

This asteroid has undergone several forms of transformation and behavior in different parts of the **Chrology** during the duration of its travel:

Suppose in the Substra **Machrology** that this asteroid begins its journey with **Ei** (Initial energy) and its initial conception (**Ci**). Before we receive this energy on earth it has undergone several modifications in its positions and in the time. If the energy will put several Light Years (**AL**) before reaching us it will no longer behave in the same way as its initial configuration (original **Ei** and **Ci**) and the different transformations in space during the time travel before we received and recognized at our position on earth is a consequence of the **Chrologic Physics**:

Ei, Ci ≠ Ef, Cf

Ef = Final energy
Cf = Final conception

Mi C + Mi ≠ Mf C + Mf
=> Mi (C + 1) ≠ Mf (C + 1)
Mi ≠ Mf

This example will be demonstrated with more precision in the scientific editions include many fact and cases in our real lives.

Chrology defines five (5) major study divisions that correspond to the five (5) worlds of study in our system throughout all existence of universe.

The **five areas of the first principle** of **Chrology** are:

- Normal Domain
- Microscopic Domain
- Macroscopic Domain
- Extra-Microscopic Domain
- Extra-Macroscopic Domain

In these five areas, any studies described by man must be classified within these distributions.

All existences in the **universe** are present in the **five (5) domains** of the **First Principle of Chorology**, but always limits compared to the other domains.

To study and to understand them originally, **Chrology** has separated them well to allow them to be accessed to us in an original way without modifying its aspects and to make their deep study.

The **five (5) areas** of study throughout the **universe** are:

1. Substranormalogy
2. Substramachrology
3. Substramichrology
4. Substraextramichrology
5. Substraextramachrology

1. Substramicrology: is the group of all the Sciences that study the natural or artificial creatures that are in the sub-micro domain in the first principle of Chrology. Every existence in the all universe called substra in sub-stramicro domain is studied in Substramicrology.

2. Substramachrology: is the group of all sciences that studies in all senses the natural and artificial creatures that are grouped in the Substramacro.

3. Substranormalogy: is the group of science literature that study movements, energies, forces, different positions, different behaviors... of natural or artificial creatures that are in the Substranorma. In Chrology many studies reside in this group where the Leugs are very advanced and numerous to us.

4. Substraextramichrology: is the study of the limit of natural creatures in Substra-extramicro. It is the natural creatures, existences, invisible to our electronic optical microscopes, which tend towards the infinitely small out of our wear to study them with our standard models.

5. Substraextramachrology: is the study of natural creatures being at an exaggeratedly large distance within the substraextramacro.

Those are creatures, existences that are invisible to our glasses, binoculars, probes, space satellites or all the instruments that allow us to see far into the horizon of the universe we now have.

It's all existences out of our wear to study them with our standard models in the **Macro**. This part deals without the ordinary problems that are disappointing to the Big Bang Theory.

Chapter 7

Fundamental Theorem of Total Mass of the Universe Described by Chrology

Content

Introduction

The overview of the whole universe on our human scale with our different materials and study instruments is an absolute complex.

But to use all our methods our techniques our instruments our discoveries that are understood to our proportion on our fields of vision, understanding and help us to determine the total mass of the whole universe paradoxically leads us to a more absurd work and consequently, several universes have been described by different thinkers, scientists, scholars, religious some of which may be mentioned with different insights and understandings:

- The Fractal universe,
- The universe of Einstein,
- The hesitant universe,
- The universe in two-dimensional torus,
- The Ekpyrotic Universe,
- The Universe of Milne,

- The Universe of Gödel,
- The Universe of Eden,
- The Mixmaster Universe,
- The Universe of De Sitter,
- The universe of Friedman-Lemaitre-Robertson-Walker, etc.

All these descriptions leave us with insufficient understanding of the entire universe, and we are completely on a very broad and very absurd method to shed light on the original description of the total mass of the entire universe.

If we focus on the explanations of theories currently present in our different scientific societies, some of which have been accepted by one group and rejected by another group, the search for evidence will be very important to build some light on the theory. Big Bang, String Theory, Inflation Theory, and some contradiction on Quantum Theory.

One thing that remains with several puzzles is the understanding of a simple and unique model to clarify the presence and existence of all that is around us, to delay us on different scale.

All of these remind us that there must be other simpler way to open our eyes and understand the total mass of the universe, for which the total energy of the universe and the senses of the world also for any existence differently around us.

Chrology is a glimpse of everything on their different worlds and different dimensions of existence.

If we are pointing to any part of its existence, these take us beyond our borders for its understanding but one thing that is very important is that the whole universe evolves on their different dimensions with times, completely different spaces for the whole system to exist and appear to us.

The Mass Concept and Its Evolution
The concept of mass has progressively evolved in scientific communities and according to different eras of our evolution of our progress on understanding the whole **cosmos**.

Going back in history from the understanding of the traditional mass from the alien mass, from **Newtonian mechanics** to **particle physics** and so on. In this concept according to the **Chrology** in its aspect we are very far from the deep understanding of the **total mass** of the **universe**.

In-depth studies of **chemistry** on the quantities of materials, new **concepts**, and confirmations of **quantum mechanics**, observations of restricted and **general relativity**, the understanding of the mass has changed a lot to the **Higgs Boson** and the **plank constant**.

In present-day scientific communities accepted **conventions** and progressions have described the universe on a large scale by the laws of **general relativity**, whose energy is involved in **gravitation** and not **mass**.

The **cosmology** from the **Big Bang** Theory considers the **cosmic horizon** of the universe as a visible part as well as the limit of the observable universe with our instruments to determine its **total mass** with consideration to the **density factor**; however, **the invisible** part of the **dark mater** remains an **elusive mystery**.

According to **Chorology**, one cannot determine the **dynamics** and **geometry** of the **Universe** without, however, precisely determining the **amount of material** that composes it on its different **areas of access**.

The **Cosmological** constant has been introduced as the surface curvature of space to intervene in the theoretical perception and construction of space. The gravitational constant has been introduced as a magnitude to represent all the fundamental characteristics of the universe in the interaction between all mass.

All these **explanations**, all these **demonstrations**, all these **appearances** and very **complex existence** of the **quantity of materials** in the absolute **universe** by **summing** the different **mass** of parts which constitute it with different forms and different compositions **confuses** us and makes us **superfluous** for a method of determination and exact understanding of all the **matter** that constitutes the whole of the **whole universe**.

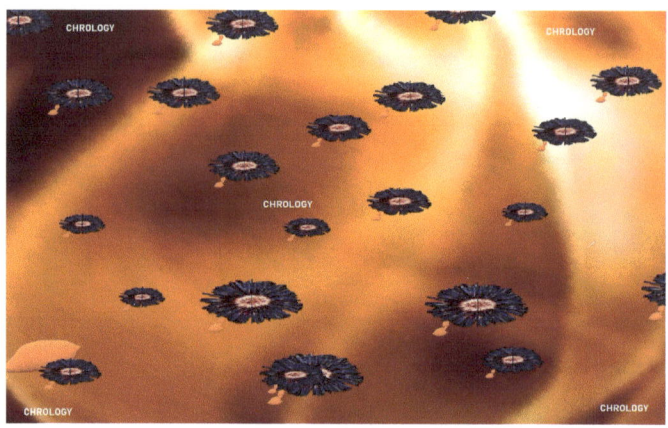

Chrology Picture 7

Determining the **total amount of matter** that constitutes the **entire cosmos** requires a new method that must consider the **entire universe** in a **single concept** and more specifically by introducing time and space in ways that are not **Chrological** and in a standard way to solve the **mystery** of the whole concept.

Several methods that have been used to describe mass have contributed to progress in **scientific communities**.

The method introduced by **Chrology** considers matter in its Chrological state in the universe before determining its **quantity** of matter which composes it the factors time and space is introduced typically in its **Chrologic aspect**.

The Total Mass of All Lunivers Demonstration

Chrology describes the mass of any body as the sum of its entire mass, which is different from the **sum** of its **quantities** of matter.

These remind us to consider a body and make **an inventory** of all its materials, in the **Chrologic division** which appears very difficult in the classical world and the sum of all these masses gives us the **total mass** of this body and all the quantities were considered even if the unit of measure must be **very large** or **very tiny**.

This apprehension and this new determination are the actual existence of a mass throughout the universe and, we are going to get new inventions to relate to the mass factors in the **Chrology** and the **calculations** with the mathematics necessitates some new parameters of the universe to introduce.

Let's take the simple example of a body **L** in one of five domains described by first **Principle of Chrology**:

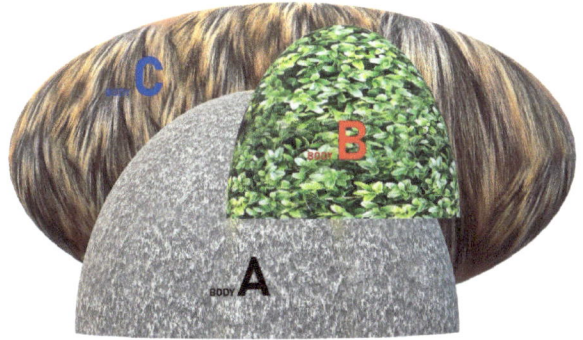

The Body L despite in **Chrology** is still composed **Chrologically** in three (3) body:

- **CA body**
- **CB body**
- **CC body**

Each Chrologic part is composed of world where matter and its parts are found, and the total sum of these masses is calculated **Chrologically**.

- **CA** Matter is the Total **CA** Mass in **L** Chrology
- The material of **CB** is the total mass of **CB** in the Chrology of L.
- The material of **CC** is the total mass of **CC** in the Chrology of **L**.

The total mass of **L** is the quantities of total matter of **L** in the different small **Chrologic world**:

$$\mathbf{M\ (Total)\ L = M\ (Total)\ CA + M\ (Total)\ CB + M\ (Total)\ CC}$$

These simple **mathematical calculations** will be developed on the **scientific** edition of **Chrology** because **anything exist** in universe present always his **Chrologic aspect**.

In a concept of the **global world** (GW) accepted as our world to generalize and calculate the total mass of the whole universe, it is the total mass of visible materials plus the total mass of dark matter because our universe is decomposed into two parts of which we know, a little about the visible universe and not much about the invisible universe that contains **dark matter** or a large part of the whole universe concentrates.

According to the first **principle of Chrology**, the absolute universe is decomposed into **five parts**, including all the worlds of measurable and visible matters, and all the worlds of non-measurable and invisible **black matter**. Every world has a visible and invisible **mass of materials**.

Chrology tells us in its principles to speak about some things in the universe it is necessary that this thing **exists** and if it exists it must be present somewhere in a well specified world, ordered with its **laws** and principles that the universe himself allows it.

In the example of the body **L** precedes its Chrology **L** in front of our position of study is the sum of its existence **(EL)** by its presence **(PL)** in a very precise domain which is equal to:

Chrology of L = EL + PL

    EL = Existence of L PL = Presence of L

    PL = Total mass of **L (ML)** time the square of the velocity(of measurement in **Chrologic domain** where L is located.

$$\text{Chrology of L} = ML + ML \cdot V^2$$

$$\text{Chrology of L} = ML\,(1 + V^2)$$

$$\text{Chrology of body L} = K. \sum_{M=A}^{M=C} M$$

=K. M (Total) L

M (Total) L = M (Total) CA + M (Total) CB + M (Total) CC

Chrology of body L = K. [M (Total) CA + M (Total) CB + M (Total) CC]

With **K** the constant of **existence** and **presence** in the domain to study.

The important factor of **Chrology** to speak of its calculations is that all **matter** must exist because the invisible or the dark matter has a well-defined explanation in **Chrology**.

If there is this **matter** it is because its presence must be somewhere and as each domain evolves with speed and **energy** in the universe the calculation of the **Chrology** is done by the sum of two which is the product of **K** by the total sum of the **masses**. We will return to these calculations in the scientific edition.

The total mass of the whole universe is the **total mass** of visible matter plus the total mass of **dark matter**.

This formula is only the total mass of the **entire universe** which is admitted in the classical calculations:

Total mass of the universe = total mass of visible materials + total mass of dark matter.

Chrology gives the total mass of the whole universe by **summing** all the masses in the **different Chrologic worlds** in its first principle of the first postulate that's must show all existence of everything in all universe:

Total mass of the universe = total mass of the Chrology

T(Mu) = T (Mc)

Chrology Picture 8

$$\text{CHROLOGIC MASSES OF ALL LUNIVERSE} = \sum_{M=-C}^{M=+C} (M).C$$

The masses (M) from **-C** to **+C** depend on which Chrologic world we are located to determine the total mass.

In the general case of the whole universe there is the constant velocity that comes with the speed of light and the speed of expansion of the whole universe.

In this theorem we consider the speed of light **C** for calculations. And **K** becomes

$$K=C^2 + 1$$

This is a general constant of presence and existence.

$$\text{CHROLOGY OF UNIVERSE} = K . \sum_{M=1}^{M=5} M$$

$$K=C^2 + 1$$

$$\text{CHROLOGY OF UNIVERSE} = (C^2 + 1) . \sum_{M=1}^{M=5} M$$

In standard configurations and calculations accepted, the **TOTAL MASS OF THE UNIVERSE** is equal to the **Total Mass of Visible Matter (MV)** plus **Total Mass of Dark Matter (MN).**

TOTAL MASS OF THE UNIVERSE =Total Mass of Visible Matter **(MV)** + Total Mass of Dark Matter **(MN)**

TOTAL MASS OF THE UNIVERSE =(MV) +(MN)

Demonstration:

Take the Chrology of the whole universe which is written:

$$\text{CHROLOGY OF UNIVERSE} = (C^2 + 1) . \sum_{M=1}^{M=5} M$$

$$(C^2 + 1) . \sum_{M=1}^{M=5} M = M1 + M1C^2 + M2 + M2C^2$$

$$+ M3 + M3C^2 + M4 + M4C^2 + M5 + M5C^2$$

$$= M1 (C^2 + 1) + M2(C^2 + 1) + M3(C^2 + 1)$$

$$+ M4(C^2 + 1) + M5(C^2 + 1)$$

$$= (C^2 + 1)(M1 + M2 + M3 + M4 + M5)$$

$$(C^2 + 1) . \sum_{M=1}^{M=5} M = (C^2 + 1)(M1 + M2 + M3 + M4 + M5)$$

By eliminating the constant of existences and presences we obtain:

$$\sum_{M=1}^{M=5} M = M1 + M2 + M3 + M4 + M5$$

$$\sum_{M=1}^{M=5} M$$

This is the sum of all the masses of matter in the universe.

WITH:

- **M1** = Total mass of all existences in the Normalogy **M (n)**
- **M2** = Total mass of all existences in Machrology **M (M)**
- **M3** = Total mass of all existences in Michrology **M (m)**
- **M4** = Total mass of all existences in Extramichrology **M (extrami)**
- **M5** = Total mass of all existences in Extramachrology **M (extraMa)**

We know that the total mass of the universe is equal to the total mass of the "Chrology"

Starting from this equality in the fundamental case of the whole universe we obtain:

$$\textbf{CHROLOGIC MASS OF ALL UNIVERSE} = \sum_{M=-C}^{M=+C} \textbf{(M)} . \textbf{C}$$

This is equal to:

$$\sum_{M=-C}^{M=+C} \textbf{(M)} . \textbf{C} = [\ \textbf{M(n)} + \textbf{M(m)} + \textbf{M(M)} + \textbf{M(extami)} + \textbf{M(extraMa)}\ ]$$

$$\sum_{M=-C}^{M=+C} \textbf{(M)} . \textbf{C} = \sum_{M=1}^{M=5} \textbf{M}$$

All equal to:

$$\sum_{M=1}^{M=5} \textbf{M} = [\ \textbf{M(n)} + \textbf{M(m)} + \textbf{M(M)} + \textbf{M(extami)}$$

$$+ \textbf{M(extraMa)}\ ]$$

The **Chrologic Mass of all Universe**= The **Total Mass of the Universe** = **Total Mass of Visible Substances** + **Total Mass of Black Substances**

With:

- Total Mass of Visible Substances = **MV**
- Total Mass of Black Substances = **MN**

$$MV + MN = \sum_{M=-C}^{M=+C} (M).C = \sum_{M=1}^{M=5} M$$

$$MV + MN = M(n) + M(m) + M(M)$$
$$+M(extami) + M(extraMa)$$

THIS WAS TO BE DEMONSTRATED.

We will return to this part on the scientific edition for more demonstrations.

Chrology deals with the problems of the foundation and structure of the skeleton of the whole universe in a simpler theory to explain several mysteries beyond us and the simple way where everything must exist and appear to us in our correct global world.

> <<*All we possess, all we are living, seeing are from another possesses that influences another configuration until all are stable for apprehension to us. We are just in irreversible positions in all.*>>Ulrich Ndilira Rotam.

Acknowledgment

Special thanks to all who involve in my life in different way, and I learned new things within the duration we spend together.

I would like to thank different places I spend times during the researches in U.S.A., France, Japan, UK, West, Africa, Italia, Switzerland, Belgium.

I am also grateful to my friends, my different colleagues from different studies, from different works places from different countries who support and believe in my high quality in physics, cosmology, mathematics, climate change solutions, complicate problem solving, and extraordinary services qualities provides in different work places.

I would like to thank my parents, whose love and always support and believe on me in different moment, I thank my loving and supportive wife, Albertine, my two wonderful <<Baby>> Millard and Hershel, who provide some inspiration indirectly to this realization and my lovely sister Bibiane Mbakass.

Special thanks to Crystal Ngoje, Christian. T, Gouara. D, Drackey. L, Sébastien. K, Benjamin. M, Dimanche. T, Sialngar. A, Konate. M, Eric. N, Armel. M, Dave. H, Anthony. C, Sayaka. K, Massa. K, Maryse. M, Neha. M, Serge. W. F, Michel. N, Michel. M, Charles. N, Mangaral. B, Ghyslaine. B, Emanuel. N, Tosso. K, Carlos. C, Aryio. O, Amiron. W, Arnaud. S, Ox. N, Malla. O, Didier. B, Clemence. A, Mbaikoula. N, Malla. O, Piet. D, Ulrich. P, Jude. F, Hughes. M, Talomady. N, Nassirhagbe. C, Berenger. M, Lawrence. J, Hamad. O, Jude. N, Gilles. D, Amina. L, Crystal. N, Maryse. M, Roger. K, Patricia. M, Ted. H, Anaclet. D, Kameldy. N, Derbew. S, Guenan. S, Solange. G, Charles. Y, Alain. M, Guealbaye. K,

Ghislain. T, Frederick. N, Willy. A, Botha. M, Ouganda. J, Nymot. G, Jelena. K, Maye. B, Daissou. B, Félicien. N, Alladoum. F, Koumi. K, Nanadoum. K, Mahmoud. Y, Guealbaye. M, Emmanuel. N, Neloum. F, Gloria. N, Harnet. W, Dhan-Valero. B, Dee. M, Beyamra. B, Madjissembaye. T, Sotinan. J, Ntouda. H, Joachim. M, Medard. B, Mirabaye. H, Jude. N, Maye. B, Namala. G, Roger. B, Koumtigue. K, Dahouta. N, Baldal. A, Nicole. N, Kadadoum. J, Kelemi. A, Succes. M, Geraldine, Medmadjal. P, Tarassoum. T, Londadjim. T, Danielle. N, Soad. D, Allog, Valdo, Nguembi, Prosper. E, Chivas, Konate, Raphael, Paulin. E, Baldal. A, Marco. T, Botha. B, Yoko. A, Koudjal. A, Patrick. M, Honore, Renald, Christophe. D, Koina. A, Malla. O, Felix. N, Joakim. M, Olivier. M, Olivier. D, Edwich. E, Fatou. L, Ndinira. N, Felix. P, Ngueto. T, Alifa, Medard, Sylvestre. K, Serge. G, Solange. G, Mirabaye. S, Neloum, Asnal. E, Odile, Tony, Camille, Junior, Stephanie, Claudia, Bernice, Nadege. D, Korna, Jimmy, Ndoude, Olivier. M, Alain, Emanuel. G, Kregoto. A, Gloria. R, Mireille. N, Grace. M, Gladys. D, Hughes. N

References

1. A. N. The Art of Living: Socratic Reflections from Plato to Foucault Sather Classical Lectures First Edition.

2. D. M. John Adams.

3. J. M. Thomas Jefferson: The Art of Power.

4. J. J. E. The Quartet: Orchestrating the Second American Revolution, 1783-1789.

5. J. J. E. Revolutionary Summer: The Birth of American Independence.

6. J. J. E. American Sphinx: The Character of Thomas Jefferson.

7. J. G. God in My Head: The true story of an ex-Christian who accidentally met God.

8. K. C. God and Hamilton: Spiritual Themes from the Life of Alexander Hamilton and the Broadway Musical He Inspired Kindle Edition.

9. R. C. Alexander Hamilton.

10. R. C. Washington: A Life.

11. S. H. M. Life After Death, Powerful Evidence You Will Never Die: Second Edition.

12. V. A. The Original Teachings of Jesus Christ.

13. M. J. C. Mechanics from Aristotle to Einstein.

14. W. H. The Best of Socrates: The Founding Philosophies of Ethics, Virtues & Life 3rd Edition.

15. (Editor), by M. I. F. The Portable Greek Historians: The Essence of Herodotus, Thucydides, Xenophon, Polybius (Viking Portable Library) 1st Edition.

16. 1 & BARASH, E. B. J. A. The Symbolic Construction of Reality THE LEGACY OF ERNST CASSIRER.

17. A. MAZURE (Ed), Alain Blanchard, Françoise Combes, and P. B. Galaxies and Cosmology.

18. Abbott, A. E. V. D. and I. Theory of Wing Sections: Including a Summary of Airfoil Data.

19. Adams, B. by R. A. Calculus: A Complete Course.

20. Ahmad, I. Mathematical Study of Multi-Fields Inflation and Perturbation Theory.

21. Alan Lothian, Michael Kerrigan, P. V. Epics of Early Civilization: Middle Eastern Myth.

22. Alexander, by B. S. R. and T. D. New Dictionary of Biblical Theology: Exploring the Unity Diversity of Scripture.

23. Allan, S. M. and T. Ancient Greece and Rome: Myths and Beliefs.

24. Allan, T. L. and T. Gods of Sun and Sacrifice: Aztec and Maya Myth.

25. Allan, T. Exploring the Life, Myth, and Art of the Medieval World.

26. Allan, T. Exploring the Life, Myth, and Art of South America.

27. Allan, T. Issues Today.

28. Allan, T. Ancient China.

29. Allan, T. The Roman World.

30. Allan, T. Exploring the Life, Myth, and Art of Japan.

31. Allen, J. As a Man Thinketh.

32. Allen, P. Mr. Archimedes' Bath.

33. Almqvist, E. History of Industrial Gases.

34. Altintas, Z. Biosensors and Nanotechnology: Applications in Health Care Diagnostics.

35. Amdahl, K. There Are No Electrons: Electronics for Earthlings.

36. Anderson, J. Mechanically Inclined.

37. Anderson, M. Isaac Newton.

38. Andrew Fox, R. D. Gas Accretion onto Galaxies.

39. Angus, D. Great Inventors and Their Inventions.

40. Archimedes. The Works of Archimedes: Volume 2, On Spirals.

41. Archimedes. The Method of Archimedes, Recently Discovered by Heiberg; A Supplement to the Works of Archimedes, 1897.

42. Archimedes. The Works of Archimedes.

43. Aristotle. The Philosophy of Aristotle.

44. Aristotle. Poetics.

45. Aristotle, B. by. Politics.

46. Aristotle, B. by. Nicomachean Ethics.

47. Arroyo, S. Astrology, Psychology and the Four Elements.

48. Asimov, I. Beginnings: The Story of Origins—of Mankind, Life, the Earth, the Universe.

49. Asimov, I. Fantastic Voyage II : Destination Brain.

50. Asimov, I. Asimov's Chronology of the World.

51. Austen, J. Pride and Prejudice.

52. Author), by J. P. (Goodreads. The Passion of Jesus Christ.

53. Bavinck, by H. Reformed Dogmatic: Abridged in One Volume.

54. Beard, M. Women & Power.

55. Becker, A. What is Real?

56. Becker, R. O. The Body Electric.

57. Bendick B. (author) J. Archimedes and the Door of Science.

58. Bettison, I. Edexcel AS and A Level Mathematics Statistics & Mechanics Year 1/AS Textbook + e-book.

59. Birch, by B. Marie Curie's Search for Radium.

60. Bishop, R. L. Tensor Analysis on Manifolds.

61. Blakey, R. History of the Philosophy of Mind.

62. Blakey, R. History of the Philosophy of Mind, Embracing the Opinions of All Writers on Mental Science from the Earliest Period to the Present Time.

63. Boland, Y. Astrology.

64. Boljanovic, V. Applied Mathematical and Physical Formulas.

65. Books, T.-L. The Diamond Path: Tibetan and Mongolian Myth.

66. Books., N. G. Planets, Stars, and Galaxies: A Visual Encyclopedia of Our Universe.

67. Borisenko, B. (author) A. I. Vector and Tensor Analysis with Applications. in

68. Born, B. by A. E. and M. The Born-Einstein Letters.

69. Born, M. Atomic Physics.

70. Boyd, R. How humans evolved.

71. Brandt, S. A. Introduction to Aeronautics: A Design Perspective.

72.Bremness L. Herbs.

73.Louis de Broglie. La Physique nouvelle et les quanta.

74.Louis de Broglie.Matter and Light - The New Physics.

75.Louis de Broglie.Un itinéraire scientifique (Histoire des sciences) (French Edition).

76.Louis de Broglie.Jalons pour une nouvelle microphysique : Exposé d'ensemble sur l'interprétation de la mécanique ondulatoire.

77.Louis de Broglie.Nouvelles perspectives en microphysique.

78.Louis de Broglie.Ondes et mouvements (Les Grands classiques Gauthier-Villars) (French Edition).

79.Louis de Broglie.Ondes corpuscules mécanique ondulatoire.

80.Louis de Broglie. La Physique nouvelle et les quanta.

81.Brydson, R. Electron Energy Loss Spectroscopy.

82.Bryson, B. A Short History of Nearly Everything.

83.Bryson, B. A Really Short History of Nearly Everything.

84.Bunn, J. H. The Natural Law of Cycles.

85.By (author) A. R. Hibbs, B. (author) R. P. F. Quantum Mechanics and Path Integrals.

86.(Author) David Lovelock, B. (Author) H. R. Tensors, Differential Forms and Variational Principles.

87.(Author) Gennaro Auletta, (Author) Giorgio Parisi, B. (Author) M. F. Quantum Mechanics.

88.By (author) Matthew Sands, By (author) Richard P. Feynman, B. (author) R. B. L. The Feynman Lectures on Physics, Vol. III: The New Millennium Edition: Quantum Mechanics.

89.(Author) Piers Bizony, F. by J. A.-K. Atom (Icon Science).

90.(Author) R.H. Barnard, (Author) D.R. Philpott, B. (author) A. C. K. Mechanics of Flight.

91.By Simmone Hewett, By Sue Holt, By Keith Johnson, B. J. M. Advanced Physics for You.

92.By (author) Sunquist Irvin, By (author) Scott W Sunquist, B. (author) D.

T. I. History of the World Christian Movement.

93. Benjamin T. Solomon. Super Physics for Super Technologies: Replacing Bohr, Heisenberg, Schrodinger and Einstein.

94. Erwin Schrodinger, R. P. What is Life? With Mind and Matter and Autobiographical Sketches / Edition 1.

95. F. David Peat, Danah Zohar, I. M. Who's Afraid of Schrodinger's Cat: The New Science Revealed - Quantum Theory, Relativity, Chaos and the New Cosmology.

96. Leah Ceccarelli. Shaping Science with Rhetoric: Cases of Dobzhansky, Schrodinger, and Wilson - 01 edition.

97. By Michael Walzer (Editor), M. R. (Translator). Regicide and Revolution: Speeches at the Trial of Louis XVI.

98. Paul Dowswell Ruth Brocklehurst, H. B. The World Wars: An Introduction to the First and Second World Wars.

99. Carr N. The Shallows: What the Internet Is Doing to Our Brains.

100. Carroll L. Best Stories of All Time.

101. Carter D. B. W. and C. B. Transmission Electron Microscopy.

102. Chamberlain M. E. The Scramble for Africa.

103. Chesterton G. K. Orthodoxy.

104. ChopraD. Seven Spiritual Laws of Success: A Practical Guide to the Fulfillment of Your Dreams.

105. Chopra D. Creating Affluence.

106. Churchill W. The Story of the Malakand Field Force.

107. Churchill W. The Grand Alliance: The Second World War.

108. Churchill W. Savrola.

109. Churchill W. Arms and the Covenant.

110. Cicero. Tusculanae Disputationes.

111. Clason G. S. The Richest Man in Babylon.

112. Bill Clinton. Putting People First: How We Can All Change America.

113. Clutton-Brock, J. Mammals.

114. COLE, A. The Birth of Theory.

115. Connor, J. Kepler's Witch.

116. Courant, B. by H. R. and R. What Is Mathematics? in

117. Cox, B. Human Universe.

118. Crowe, M. J. A History of Vector Analysis: The Evolution of the Idea of a Vectorial System.

119. Daron Acemoglu, J. A. R. Why Nations Fail.

120. Davies, N. Tiny Creatures.

121. Day, B. by J. S. and T. NBiocalculus: Calculus for Life Sciences.

122. Demosthenes. The oration of Demosthenes on the crown.

123. Demosthenes. Demosthenes, Speeches 1–17.

124. Denny M. Super Structures.

125. DeVorkin D. H. Hubble imaging space and time.

126. DeVorkin D. H. Hubble.

127. Dick, P. K. The Man in the High Castle.

128. Dickinson, B. (author) T. Hubble's Universe.

129. Dickinson, T. Hubble's Universe.

130. Dolnick, E. The Clockwork Universe.

131. Draine B. T. Physics of the Interstellar and Intergalactic Medium.

132. Drexel A. MENTAL TOUGHNESS: Develop an Unbeatable Mind.

133. Ebert, J. D. Giant Humans, Tiny Worlds.

134. Edited by Andrew Cameron, E. by B. S. R. Trials of Theology: Becoming a proven worker in a dangerous business.

135. Albert Einstein. Cosmic Religion: With Other Opinions and Aphorisms.

136. Albert Einstein, 1879-1955. Relativity: The Special and the General Theory.

137. Albert Einstein. The World as I See It.

138. Albert, Einstein. Geometries und Erfahrung.

139. Albert Einstein. A Stubbornly Persistent Illusion.

140. Albert Einstein. Bite-size Einstein.

141. Albert Einstein. Investigations on the theory of the Brownian movement.

142. Albert Einstein. The principle of relativity.

143.Albert Einstein. The Cosmic View of Albert Einstein: Writings on Art, Science, and Peace.

144.Eisenhower D. D. Peace with Justice.

145.Ellenberg J. How Not to Be Wrong.

146.Ellis B. T. Question Everything.

147.Erickson M. J. Christian Theology.

148.Felber A. Schrodinger's Ball.

149.Feynman, R. The Pleasure of Finding Things Out: The Best Short Works of Richard Feynman.

150.Richard Feynman. Don't You Have Time to Think?

151.Fleming, A. L. and F. Egyptian Myth: The Way to Eternity.

152.Fleming, C. P. and F. Voices of the Ancestors: African Myth.

153.Frederick Douglass, Harriet Ann Jacobs, Sojourner Truth, and S. N. Voices of Freedom: Four Classic Slave Narratives.

154.Gallagher, T. F. Rydberg Atoms (Cambridge Monographs on Atomic, Molecular and Chemical Physics).

155.Gandhi. M. All Men Are Brothers.

156.Gandhi. M. Non-Violent Résistance.

157.Gandhi, M. Pathway to God.

158.Gandhi, M. Constructive Program Its Meaning and Place.

159.Gandhi, M. The Mind of Mahatma Gandhi.

160.Garrett, G. Solids, Liquids, and Gases.

161.Gater W. The Cosmic Keyhole: How Astronomy Is Unlocking the Secrets of the Universe.

162.Gianopolous A. Isaac Newton and the Laws of Motion.

163.Gibbons, G. Galaxies, Galaxies!

164.Giglio, L. Goliath Must Fall Study Guide: Winning the Battle Against Your Giants.

165.Gilligan, S. Generative Trance: The Experience of Creative Flow.

166.Gillispie C. C. Pierre-Simon Laplace, 1749-1827.

167.Gleick J. Genius: The Life and Science of Richard Feynman.

168.Gleick J. Chaos.

169.Golden Deer Classics (Author), Sigmund Freud (Author), Musashi Miyamoto (Author), Sun Tzu (Author), Voltaire (Author), H. G. Wells (Author), V. (Author). 30 Human Science Masterpieces You Must Read Before You Die.

170.Goldsmith. B. Obsessive Genius: The Inner World of Marie Curie (Great Discoveries Series).

171.Graham. B. The Intelligent Investor.

172.Green. R. Classical Theories of Money, Output and Inflation.

173.Greengrass. M. Governing passions.

174.Gribbin. J. In Search of Schrodinger's Cat: Quantum Physics and Reality.

175.Gribbin. J. Schrodinger's Kittens and the Search for Reality: Solving the Quantum Mysteries.

176.Grudem. W. Systematic Theology.

177.GUERY-ODELIN, C. C.-T. and D. ADVANCES IN ATOMIC PHYSICS: AN OVERVIEW.

178.Guinot, C. A. B. The Measurement of Time: Time, Frequency and the Atomic Clock.

179.Gulen, M. F. Selected Prayers of Prophet Muhammad.

180.Guthrie, W. General History of the World, from the Creation to the Present Time.

181.Hadfield, C. The Darkest Dark.

182.Hall, J. The Astrology Bible.

183.Hamill, P. A Student's Guide to Lagrangians and Hamiltonians.

184.Hardback, A. Motion & Kinematic.

185.Harman, P. M. Energy, Force and Matter: The Conceptual Development of Nineteenth-Century Physics.

186.Harman, P. M. Energy, Force and Matter: The Conceptual Development of Nineteenth-Century Physics.

187.Harman, P. M. The Investigation of Difficult Things: Essays on Newton and the History of the Exact Sciences in Honor of D. T. Whiteside.

188.Harman, P. M. The natural philosophy of James Clerk Maxwell.

189.Harman, P. M. Energy, Force and Matter: The Conceptual Development of Nineteenth-Century Physics (Cambridge Studies in the History of Science).

190.Hartog, J. P. Den. Mechanics.

191.Harvey-Smith, L. When Galaxies Collide.

192.Hawken, P. The Ecology of Commerce: A Declaration of Sustainability.

193.Hawking, S. Black Holes and Baby Universes and Other Essays.

194.Herzberg, G. Atomic Spectra and Atomic Structure.

195.Hickey, I. M. Astrology, a Cosmic Science.

196.Hill, J. A. Medical Astrology.

197.Hill, N. Think and Grow Rich.

198.Hill, P. G. Mechanics and thermodynamics of propulsion.

199.Hinrichs, K. Energy Its Use and the Environment.

200.Adolf Hitler. Hitler's Table Talk.

201.Adolf Hitler. Hitler's Table Talk.

202.Adolf Hitler. Mein Kampf.

203.Hoffman, A. J. How Culture Shapes the Climate Change Debate.

204.Hoffman, P. Archimedes' Revenge.

205.Hoffmann, B. by B. Albert Einstein: Creator and Rebel.

206.Infeld. A. E. and L. The Evolution of Physics.

207.Irwin, W. The Big Bang Theory and Philosophy.

208.Isaacson, W. Leonardo Da Vinci.

209.Ivanoff, V. Engineering Mechanics.

210.Johnstone, L. The Ultimate Guide to Your Microscope.

211.Jr. H. E. N. Beyond the Atmosphere: Early Years of Space Science.

212.Jr. Martin Luther King. Strength to Love.

213.Jr. Martin Luther King. The Trumpet of Conscience.

214.Kakkar, R. Atomic and Molecular Spectroscopy: Basic Concepts and Applications.

215.Kaku, M. The Future of Humanity: Terraforming Mars, Interstellar Travel, Immortality, and Our Destiny Beyond Earth.

216.Kaku, M. Hyperspace.

217. John Fitzgerald Kennedy. The strategy of peace.

218. John Fitzgerald Kennedy. President Kennedy Speaks.

219. Kennedy, P. The Rise and Fall of the Great Powers: Economic Change and Military Conflict from 1500 to 2000.

220. Kerr, P. Dark Matter.

221. Kimball, D. B. and D. Atomic physics: An exploration through problems and solutions.

222. Kimball, D. B. and D. F. J. Optical Magnetometry.

223. Kingsley, C. The limits of exact science as applied to history.

224. Knecht, R. The French wars of religion, 1559-1598.

225. Knight, R. D. Physics for Scientists and Engineers.

226. Kornfield, J. Teachings of the Buddha.

227. Kranz, G. Failure Is Not an Option.

228. Lama, D. The Universe in a Single Atom: The Convergence of Science and Spirituality.

229. LAMPERT, L. How Philosophy Became Socratic.

230. Lanczos C. The Variational Principles of Mechanics.

231. Lanczos C. The Variational Principles of Mechanics.

232. Larsen, C. Our Origins: Discovering Physical Anthropology.

233. Larsen, C. S. Bioarchaeology: Interpreting Behavior from the Human Skeleton.

234. Le, K. K. and T. First Aid for the Basic Sciences, General Principles.

235. Leonhard Euler (Author), S. H. (Editor). Elements of Algebra.

236. Levin, H. Why literary criticism is not an exact science.

237. LEVINE, E. J. Dreamland of Humanists WARBURG, CASSIRER, PA-NOFSKY, AND THE HAMBURG SCHOOL.

238. Lewis, J. L. and W. Teaming with Microbes: The Organic Gardener's Guide to the Soil Food Web.

239. Liepmann, H. W. Elements of Gas dynamics.

240. Lifshitz, E. M. Mechanics.

241. Lincoff Gary, T. L. The Knopf Mushroom Book: How to Identify, Gather,

and Cook Wild Mushrooms and Other Fungi.

242. Abraham Lincoln. Lincoln on Race and Slavery.

243. Abraham Lincoln. Lincoln Speeches.

244. Lindauer, J. The General Theories of Inflation, Unemployment, and Government Deficits.

245. Links D. W. University Math and Statistics.

246. Links D. W. University chemistry and biochemistry.

247. Links D. W. University Math and Statistics.

248. Livesey D. L. Atomic and Nuclear Physics (A Blaisdell book in the pure and applied sciences).

249. Lloyd G. The Social Pandemic: The Influence and Effect of Social Media on Modern Life.

250. Losure M. Isaac the Alchemist: Secrets of Isaac Newton, Reveal'd.

251. Louis De Broglie, Alwyn van der Merwe, G. Lochak, P. B. Heisenberg's Uncertainties and the Probabilistic Interpretation of Wave Mechanics: with Critical Notes of the Author (Fundamental Theories of Physics).

252. Louis de Broglie, Georges Lochak, Michel Karatchentzeff D. F. & Hardcover. Diverses questions de mécanique et de thermodynamique classiques et relativistes.

253. Louis de Broglie. Ondes corpuscules mécanique ondulatoire.

254. Louis de Broglie. P. The Revolution in Physics: A Non-Mathematical Survey of Quanta New edition of 1953.

255. Lowe, C. N. and S. Cosmos: The Infographic Book of Space.

256. Lupisella, E. M. Cosmos & Culture: Cultural Evolution in a Cosmic Context.

257. Lüth, by H. Quantum Physics in the Nanoworld: Schrodinger's Cat and the Dwarfs.

258. Lynn Kilgore, Robert Jurmain and W. T. Essentials of physical anthropology.

259. Macalister, T. Einstein's God: A Way of Being Spiritual Without the Supernatural.

260. Machiavelli, N. The Prince.

261. Major, F. G. The Quantum Beat: Principles and Applications of Atomic Clocks.

262. Maltz M. Psycho-Cybernetics Updated and Expanded.

263. Mandal, A. K. Electron Microscopy of the Kidney: In Renal Disease and Hypertension: A Clinicopathological Approach.

264. Nelson Mandela. No Easy Walk to Freedom.

265. Nelson Mandela. Nelson Mandela's Favorite African. Folktales.

266. Mankiw, G. Principles of Economics: Mankiw's Principles of Economics.

267. Mann, Theodore G. and N. The Elements: A Visual Exploration of Every Known Atom in the Universe.

268. Manz, C. C. The Power of Failure.

269. Martel, G. The Origins of the First World War.

270. Martin, B. R. Nuclear and Particle Physics: An Introduction.

271. Martin, B. R. Statistics for Physicists.

272. Martin, R. D. Primate origins and evolution.

273. Martin, T. C. The inventions, researches and writings of Nikola Tesla.

274. Masi, M. Quantum Physics: an overview of a weird world: A primer on the conceptual foundations of quantum physics for all.

275. Maslow, M. I. and K. Tiny World Terrariums: A Step-by-Step Guide to Easily Contained Life.

276. Matteucci F. The chemical evolution of the galaxy.

277. James Maxwell. A Treatise on Electricity and Magnetism.

278. Maynard, C. Micro Monsters: Life Under the Microscope.

279. McCauley, M. The Origins of the Cold War, 1941-1949.

280. McCoy K. Spiritual Astrology.

281. McCullough D. The Wright Brothers.

282. McPherran M. L. The Religion of Socrates.

283. McPherson J. The Ocean of Truth.

284. McTaggart L. The Field.

285. Merrifield J. B. and M. Galactic Astronomy.

286. Metcalf P. van der S. and H. Atoms and Molecules Interacting with Light: Atomic Physics for the Laser Era.

287. Miranda J. M. and L.-M. Hamilton: The Revolution.

288. Morgan C. F. A. and E. B. Supermarine Aircraft Since 1914.

289. Moss, I. G. Quantum Theory, Black Holes and Inflation.

290. Nagasawa, M. Schrödinger Equations and Diffusion Theory.

291. Nataraj N. Earth and Space: Photographs from the Archives of NASA.

292. Nataraj N. The Planets: Photographs from the Archives of NASA.

293. Nataraj N. The Planets.

294. Needle C. Tiny Worlds: Creative Macrophotography Skills.

295. Needle C. Tiny Worlds.

296. Neugebauer, O. E. The exact sciences in antiquity.

297. Neuhauser C. Calculus for biology and medicine.

298. Newbury, J. G. and D. E. Scanning Electron Microscopy and X-Ray Microanalysis.

299. Newton, I. The System of the World.

300. Newton, S. I. A Historical Account of Two Notable Corruptions of Scripture.

301. Newton, S. I. The Chronology of Ancient Kingdoms.

302. Newton, S. I. Philosophiae Naturalis Principia Mathematica (1822).

303. Newton, S. I. Newton's Philosophy of Nature.

304. Newton, S. I. The Principia: The Authoritative Translation.

305. Newton, S. I. Optics.

306. Newton, S. I. The Principia.

307. Noble, B. &. Systematics as Cyberscience: Computers, Change, and Continuity in Science.

308. Noble, B. &. Practical Cyber Intelligence.

309. Noble, B. &. Data Science for Cyber-security.

310. Barack Obama. Change We Can Believe In.

311. Barack Obama. The Audacity of Hope: Thoughts on Reclaiming the American Dream.

312. Ody P. The Complete Medicinal Herbal.

313. Oehry A. F. M. and B. P. Radiation Trapping in Atomic Vapors.

314. Ohillips C. P. and C. Aztec & Maya.

315. O'Neil C. Weapons of Math Destruction.

316. Overy R. The Origins of the Second World War.

317. Packer. J. I. Knowing God.

318. Pais. A. Subtle is the Lord...: The Science and Life of Albert Einstein.

319. PANGLE, L. S. Virtue Is Knowledge THE MORAL FOUNDATIONS OF SOCRATIC POLITICAL PHILOSOPHY.

320. Payne, E. The Pharaohs of Ancient Egypt.

321. Peery, D. J. Aircraft structures.

322. Pellant. C. Rocks and Minerals.

323. Pember. C. C. and D. R. Mass Media Law.

324. Petersen, C. C. Astronomy 101.

325. Petersen, C. C. Space Exploration.

326. Petersen, C. C. Visions of the Cosmos.

327. Petzold C. Code.

328. Pfiffikus. Protons, Neutrons, Electrons and Quarks! Tiny Atoms We Can't See.

329. Max Planck. Treatise on thermodynamics.

330. Max Planck. The Theory of Heat Radiation.

331. Max Planck. A Survey of Physical Theory: (formerly Titled: A Survey of Physics).

332. Max Planck. Where is science going.

333. Max Planck. The Origin and Development of the Quantum Theory.

334. Max Planck. The Origin and Development of the Quantum Theory.

335. Max Planck. Eight lectures on theoretical physics, delivered at Columbia University in 1909.

336. Max Planck. Between Elite and Mass Education: Education in the Federal Republic of Germany.

337. Max Planck. The universe in the light of modern physics.

338. Max Planck. Introduction to theoretical physics.

339. Plato. Republic.

340. Plato. Great Dialogues of Plato.

341. Potter M. Fluid Mechanics Demystified.

342. Ronald Reagan. The Evil Empire Speech, 1983.

343. Ronald Reagan. The Quest for Peace, the Cause of Freedom.

344. Ronald Reagan. Abortion and the Conscience of the Nation.

345. Redniss. L. Radioactive: Marie and Pierre Curie: Tale of Love and Fallout - 11 edition.

346. Reed, S. J. B. Electron Microprobe Analysis and Scanning Electron Microscopy in Geology.

347. Rees, M. Universe.

348. Reining, R. M. M. and L. Interacting Electrons: Theory and Computational Approaches.

349. Rhodes, R. The Making of the Atomic Bomb.

350. Richardson, H. Dinosaurs and Prehistoric Life.

351. Ridpath, I. Stars and Planets.

352. Ridpath, I. Stars and Planets.

353. Ritchie, D. M. C Programming Language.

354. Romm, J. Climate Change.

355. Rooney, A. Computer Science and IT: Investigating a Cyber Attack.

356. Rose, P. L. Heisenberg and the Nazi Atomic Bomb Project, 1939-1945: A Study in German Culture.

357. ROSEN, S. The Idea of Hegel's Science of Logic.

358. Rovelli, C. Reality Is Not What It Seems.

359. Rovelli, C. The Order of Time.

360. Rubakov, V. A. Introduction to the Theory of the Early Universe: Hot Big Bang Theory.

361. Rudin. W. Principles of mathematical analysis.

362. Russell B. The Principles of Mathematics.

363. Ryan, L. Advanced Chemistry for You.

364. Sagan, C. Cosmos: A Personal Voyage.

365. Sakmann, by K. Many-Body Schrödinger Dynamics of Bose-Einstein Condensates.

366. Sanger, D. E. The Perfect Weapon War, Sabotage, and Fear in the Cyber Age.

367. Schilling, G. Atlas of Astronomical Discoveries.

368. Schilling, G. Atlas of Astronomical Discoveries.

369. Schilling, G. Beyond the Solar System to the End of the Universe and the Beginning of Time.

370. Schutz, B. F. A First Course in General Relativity.

371. Scientist, B. (author) N. Why the Universe Exists: How particle physics unlocks the secrets of everything.

372. Scientist, B. (author) N. Where the Universe Came from: How Einstein's relativity unlocks the past, present and future of the cosmos.

373. Scientist, B. N. How Numbers Work.

374. Scientist, B. N. Why the Universe Exists.

375. Scientist, N. How Your Brain Works.

376. Scientist, N. How Evolution Explains Everything About Life.

377. Scientist N. Human Origins.

378. Scudder, P. H. Electron Flow in Organic Chemistry: A Decision-Based Guide to Organic Mechanisms.

379. Seigar, M. S. Spiral Structure in Galaxies.

380. Shapiro, B. A. Clinical application of blood gases.

381. Sharda, R. Business Intelligence, Analytics, and Data Science.

382. Shaw, B. R. M. and G. P. Particle Physics.

383. Shaw, G. J. The Pharaoh: Life at Court and on Campaign.

384. Shubin, by F. A. B. and M. A. The Schrödinger Equation.

385. Silva, by M. New International Dictionary of New Testament Theology and Exegesis Set.

386. Silve, D. De. An Introduction to The New Testament.

387. Simon, S. Stars.

388. Smil, V. Energy.

389. Smith, R. W. The Hubble Cosmos.

390. Smith, W. Pharaoh.

391. Sobel, D. Galileo's Daughter: A Historical Memoir of Science, Faith, and Love.

392. Sochi, D. T. Tensor Calculus Made Simple.

393. Sochi, T. Introduction to Differential Geometry of Space Curves and Surfaces: Differential Geometry of Curves and Surfaces.

394. Marie Curie Steele. P: The First Great Woman Scientist.

395. Stewart. J. Calculus: Early Transcendentals.

396. Strang. G. Introduction to Linear Algebra.

397. Stransky, P. H. and T. F. God on Our Minds.

398. Susskind, L. Quantum Mechanics: The Theoretical Minimum.

399. Tahir-ul-Qadri, D. M. Muhammad The Merciful - Islam on Mercy & Compassion.

400. Taylor, J. R. Classical Mechanics. in

401. Teitel, A. S. Breaking the Chains of Gravity: The Story of Spaceflight Before NASA.

402. Theodore E. Brown (Author), H. Eugene LeMay (Author), Bruce E. Bursten (Author), Catherine Murphy (Author), Patrick Woodward (Author), M. E. S. (Author). Chemistry: The Central Science.

403. TILCSIK, C. C. and A. Meltdown WHY OUR SYSTEMS FAIL AND WHAT WE CAN DO ABOUT IT.

404. Time, N. History of Aviation.

405. Time, T. E. of. Big Book of Science.

406. Tompkins, S. Aspects in Astrology.

407. Tony Allan, Charles Phillips, F. F. African Myths and Beliefs.

408. Trismegistus, H. The Emerald Tablet of Hermes.

409. Truesdell, C. Kinematics of Vorticity.

410. Truth, S. Narrative of Sojourner Truth.

411. Various. Indian Myth and Mankind the Eternal Cycle.

412. Version, K. J. The Greatest Stories of All Time.

413. Vlastos, by G. Socrates, ironist and moral philosopher.

414. Wade, N. Science Times Book of Mammals.

415. Wade, N. 'Science Times' Book of Insects.

416. Walton, D. J. H. The Lost World of Genesis One: Ancient Cosmology and the Origins Debate.

417. George Washington. The journal of Major George Washington.

418. Waterfield, R. Why Socrates Died.

419. Wattles, W. D. The Science of Getting Rich: Attracting Financial Success Through Creative Thought.

420. Wearing, J. Bacteria.

421. Wenham, G. Exploring the Old Testament: Pentateuch Vol 1: The Pentateuch.

422. Westerfeld, S. Goliath.

423. Williams, C. B. C. and D. B. Transmission Electron Microscopy: Diffraction, Imaging, and Spectrometry.

424. Williams, G. Advanced Biology for You.

425. Wilson, S. Electron Correlation in Molecules.

426. Wolf, W. The Origin and Nature of Our Institutional Models.

427. Wordery. Cyber Security.

428. Yang, J. H. H. and F. Modern Atomic and Nuclear Physics.

429. Yukimura, M. Planets.

430. Zobel, D. H. The Science of Tv's the Big Bang Theory.

431. NASA, National Geography, Discovery Sciences.

432. BBC, CNN, MSNBC, RFI, FOX NEW, TV5, MBC, CCTV, BOING, AIRBUS, MICROSOFT.